W0254359

Materials Handling in Pyrometallurgy

Titles of Related Interest—

Ashby ENGINEERING MATERIALS 1
Ashby ENGINEERING MATERIALS 2
Brook IMPACT OF NON-DESTRUCTIVE TESTING
Koppel AUTOMATION IN MINING, MINERAL AND METAL PROCESSING 1989
Ruhle METAL-CERAMIC INTERFACES
Taya METAL MATRIX COMPOSITES

Other CIM Proceedings Published by Pergamon

Bergman FERROUS AND NON-FERROUS ALLOY PROCESSES
Bickert REDUCTION AND CASTING OF ALUMINUM
Bouchard PRODUCTION, REFINING, FABRICATION AND RECYCLING OF LIGHT METALS
Chalkley TAILING AND EFFLUENT MANAGEMENT
Closset PRODUCTION AND ELECTROLYSIS OF LIGHT METALS
Dobby PROCESSING OF COMPLEX ORES
Embury HIGH TEMPERATURE OXIDATION AND SULPHIDATION PROCESSES
Jaeck PRIMARY AND SECONDARY LEAD PROCESSING
Jonas DIRECT ROLLING AND HOT CHARGING OF STRAND CAST BILLETS
Kachaniwsky IMPACT OF OXYGEN ON THE PRODUCTIVITY OF NON-FERROUS METALLURGICAL PROCESSES
Lait F. WEINBERG INTERNATIONAL SYMPOSIUM ON SOLIDIFICATION PROCESSING
Macmillan QUALITY AND PROCESS CONTROL IN REDUCTION AND CASTING OF ALUMINUM AND OTHER LIGHT METALS
Mostaghaci PROCESSING OF CERAMIC AND METAL MATRIX COMPOSITES
Plumpton PRODUCTION AND PROCESSING OF FINE PARTICLES
Purdy FUNDAMENTALS AND APPLICATIONS OF TERNARY DIFFUSION
Rigaud ADVANCES IN REFRACTORIES FOR THE METALLURGICAL INDUSTRIES
Ruddle ACCELERATED COOLING OF ROLLED STEEL
Salter GOLD METALLURGY
Thompson COMPUTER SOFTWARE IN CHEMICAL AND EXTRACTIVE METALLURGY
Twigge-Molecey PROCESS GAS HANDLING AND CLEANING
Tyson FRACTURE MECHANICS
Wilkinson ADVANCED STRUCTURAL MATERIALS

Related Journals

(Free sample copies available upon request)

ACTA METALLURGICA
CANADIAN METALLURGICAL QUARTERLY
MATERIALS RESEARCH BULLETIN
MINERALS ENGINEERING
SCRIPTA METALLURGICA

PROCEEDINGS OF THE INTERNATIONAL SYMPOSIUM ON MATERIALS HANDLING IN PYROMETALLURGY
HAMILTON, ONTARIO, AUGUST 26-30, 1990

Materials Handling in Pyrometallurgy

Editors

C. Twigge-Molecey
Hatch Associates Ltd.
Toronto, Ontario

T. Price
Inco Metals Company
Copper Cliff, Ontario

Symposium organized by Non-ferrous Pyrometallurgy Section of The Metallurgical Society of CIM

29th ANNUAL CONFERENCE OF METALLURGISTS OF CIM
29e CONFÉRENCE ANNUELLE DES MÉTALLURGISTES DE L'ICM

Pergamon Press
Member of Maxwell Macmillan Pergamon Publishing Corporation
New York Oxford Beijing Frankfurt São Paulo Sydney Tokyo Toronto

Pergamon Press Offices:

U.S.A.	Pergamon Press, Inc., Maxwell House, Fairview Park, Elmsford, New York 10523, U.S.A.
U.K.	Pergamon Press plc, Headington Hill Hall, Oxford OX3 0BW, England
PEOPLE'S REPUBLIC OF CHINA	Pergamon Press, 0909 China World Tower, No. 1 Jian Guo Men Wai Avenue, Beijing 1000004, People's Republic of China
FEDERAL REPUBLIC OF GERMANY	Pergamon Press GmbH, Hammerweg 6, D-6242 Kronberg, Federal Republic of Germany
BRAZIL	Pergamon Editora Ltda, Rua Eça de Queiros, 346 CEP 04011, Paraiso, São Paulo, Brazil
AUSTRALIA	Pergamon Press Australia Pty Ltd., P.O. Box 544, Potts Point, NSW 2011, Australia
JAPAN	Pergamon Press, 8th Floor, Matsuoka Central Building, 1-7-1 Nishishinjuku, Shinjuku-ku, Tokyo 160, Japan
CANADA	Pergamon Press Canada Ltd., Suite 271, 253 College Street, Toronto, Ontario M5T 1R5 Canada

Library of Congress Cataloging in Publication Data

ISBN 0-08-040414-6

Printing: 1 2 3 4 5 6 7 8 9 Year: 0 1 2 3 4 5 6 7 8 9

Printed in the United States of America

The paper used in this publication meets the minimum requirements of American National Standard for Information Sciences-Permanence of Paper for Printed Library Materials, ANSI Z 39.48-1984

Symposium Co-Chairmen

C. Twigge-Molecey
Hatch Associates Ltd.
21 St. Clair Avenue East
Toronto, Ontario, Canada

T. Price
Inco Metals Company
General Engineering Building
Copper Cliff, Ontario, Canada

Session Chairmen

Pneumatic Injection

D. Goodwill
H.G. Engineering
Toronto, Ontario, Canada

R. Barryman
Falconbridge Ltd.
Toronto, Ontario, Canada

Bins Segregation and Blending

G. Richards
The University of British Columbia
Vancouver, British Columbia, Canada

J. Merks
Matrix Consultants
Vancouver, British Columbia, Canada

In-line Sampling and Weighing

H. Heinen
The University of Alberta
Edmonton, Alberta, Canada

F. McCaffrey
Hatch Associates Ltd.
Toronto, Ontario, Canada

Hot Metal Handling and New Methods

A. Reed
Wolfson Centre For Bulk Materials
Handling Technology
London, England

W. Imrie
Bechtel Corporation
San Francisco, California, U.S.A.

Environmental Aspects of Materials Handling

B. Bailey
Noranda Sampling Inc.
Providence, Rhode Island, U.S.A.

N. Heywood
Warren Spring Laboratory
Stevenage Herts, London, England

Preface

The nineties are expected to be a period of intense global competition for the primary non-ferrous and ferrous metals producers. There will continue to be both cost and quality pressure from overseas as well as the threat of substitution by newer materials. To meet this competition, a generation of high intensity smelting processes are now coming on line. These demand new thinking on materials handling methods, sampling methods and controls.

At the Conference of Metallurgists, it has been traditional to focus on the process reactors and unit operations, the heart of metal production. Very little attention has been paid to the materials handling systems that link the parts of the plant. For these systems, there are transportation aspects as well as sampling methods, storage and blending techniques and environmental aspects. Proper design and application of a material handling system must address all these issues, if cost effective, quality product is to be produced.

The proceedings that follow are from the International Symposium on Materials Handling in Pyrometallurgy sponsored by the Non-ferrous Pyrometallurgy Committee of The Metallurgical Society of The Canadian Institute of Mining and Metallurgy. The Symposium was organized to bring together the plant operating groups with the researchers, developers, designers and vendors to highlight problem areas and lay down the ground work for development or implementation of new approaches.

The sections are organized according to general subject areas:
- Pneumatic Injection;
- Bins Segregation and Blending;
- In-line Sampling and Weighing;
- Hot Metal Handling and New Methods;
- Environmental Aspects.

The international aspect is reflected by papers from two continents and over 60% of the presenters from outside Canada.

We would like to extend our thanks to all authors for providing the valuable exchange of ideas and information both at the meeting and in this volume. Our thanks are also due the Session Chairmen who acted as the Symposium Technical Committee, performing the difficult task of finding authors and papers of interest to the community.

Thanks are also due the Non-ferrous Pyrometallurgy Committee for their support and to Miss Wendy Mitchell for her invaluable help throughout the organizational phase and in preparing the final manuscript.

C. Twigge-Molecey
Hatch Associates Ltd.
21 St. Clair Avenue East
Toronto, Ontario, M4T 1L9

T. Price
Inco Metals Company
General Engineering Building
Copper Cliff, Ontario, P0M 1N0

May 1990

TABLE OF CONTENTS

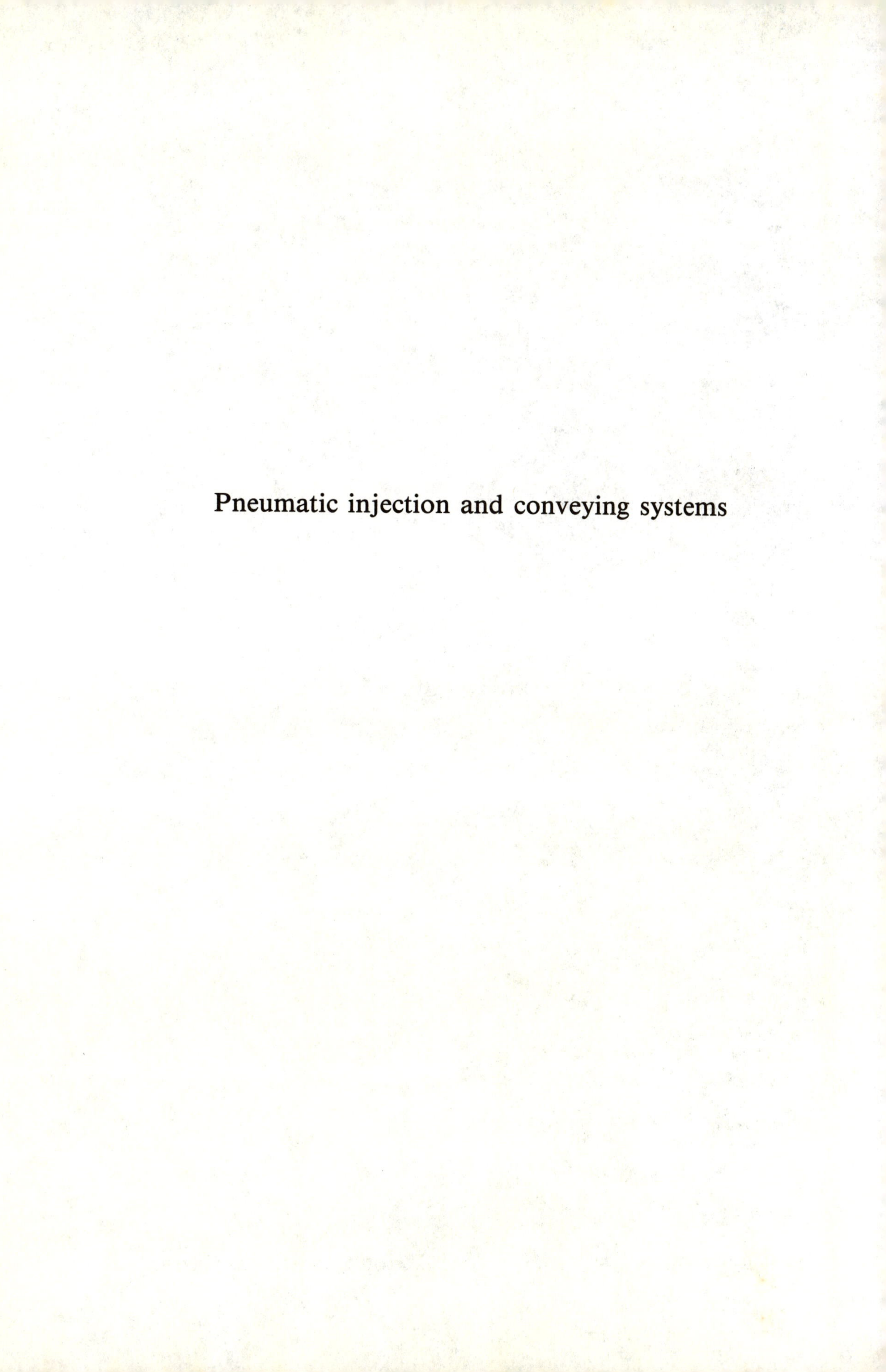
Pneumatic injection and conveying systems

Fundamental aspects of gas-particle injection

G.G. Richards
Centre for Metallurgical Process Engineering, University of British Columbia
Vancouver, British Columbia, Canada V6T 1W5

Abstract

The submerged injection of particulate material has become a important industrial process in both the non-ferrous and ferrous industries. Examples include the injection of magnesium or lime in steel processing and pulverized coal in slag fuming. This paper is a review of the fundamental aspects of the phenomena surrounding the process of solids injection. A brief summary of transport regimes of gas-solid mixtures in given followed by a discussion of the effect of the solids on the gas-liquid injection regime. The presence of the solid particles has a strong influence on whether the gas discharges as bubbles or jets as it enters the liquid. This in turn can affect accretion formation, tuyere blockage and refractory wear. Once the gas-particle jet has entered the liquid the particles can either entrain in the bath or tend to remain with the gas. This has a direct impact on the efficiency with which the solid is utilized. The major factors influencing the deportment of the solids are particle size, surface tension and particle momentum.

Introduction

The injection of particulate solids or powders with gas is a common metallurgical procedure. It is regularly used in a large number of different processes to effect chemical reactions, for submerged combustion and simply as a means to introduce solids into the bath. A number of examples are given in Table I. There are several advantages to the use of gas-particle injection including:
a) the formation of a potentially large solid-liquid contact area for reaction,
b) a potentially large solid/gas-liquid contact area for heat transfer,
c) it avoids the need to pelletize fine materials before a bath smelting step, and
d) reagents can react directly with the melt without first being diluted by the slag.

In order to effectively carry out a powder or particulate injection process attention must be paid to a number of inter-related factors. It is the objective of this paper to review the fundamental aspects of the process to serve as introduction to the complexities of the phenomena. The aim is to provide a guide for the assessment of system design parameters and as an aid in troubleshooting injection problems.

The dynamics of gas-solids injection involves a series of processes including the upstream gas-solids flow, the interaction of the liquid and gas-solids 'jet' and finally the behaviour of the solids and gas as they move away from the injection point. In dealing with these topics it is important as well to briefly review the dynamics of single particle-liquid interaction.

TABLE I. Particulate Injection Applications in Metallurgical Processing

INJECTION SYSTEM	PURPOSE	REFERENCE
Lime, calcium, calcium carbide, magnesium - steel	Dephosphorization, desulphurization and desiliconization of hot metal and steel	1-5
Coal - steel	Submerged combustion for high scrap ratios	1,6,7
Carbon - EF slag	Reduction of FeO	8
Mn - Al	Alloying	9
Coal - copper-making slag	Slag cleaning	10,11
SIROSMELT (Cu, Ni, Pb, Sn smelting)	Slag reduction/submerged combustion	12,13
Coal - Lead BF slag	Slag reduction	14
Flux - copper	Refining of anode copper	15

Pneumatic Transport and Conveying of Solids

The behaviour of a gas-particle jet on injection into a liquid is determined to a significant degree by the transport conditions of the solids and gas in the incoming stream. It is therefore important to briefly review the different flow regimes and their characteristics. The most important factor is the solids-to-gas loading. In a dilute suspension horizontal flow will be homogeneous with all solid particles suspended in the gas phase. However, as the loading increases the gas velocity is no longer sufficient to keep all of the solids suspended and the particles begin to flow as 'dunes' along the bottom of the pipe. At higher loadings the solids will move in slug flow which eventually degenerates into full plugged flow. (16,17) The precise transition between these regimes depends on a number of factors including particle size and shape, fluid and particle densities and fluid velocity. As a rule-of-thumb homogeneous flow will exist if the flow Reynolds number is greater than 10^5, the particle size is less than 200 μm and the solid-to-gas loading (wt/wt) less than 2. (16,17) The transition to non-homogeneous flow will occur when particles tend to settle or 'salt' out of suspension. The saltation velocity is the minimum fluid velocity required to carry solids at a specific rate without allowing them to settle. Based on work by Zenz (18), Wen presents the following correlation for saltation velocity, u_s, in horizontal transport:

$$u_s = a_1 D^{0.5}[4g\mu_f(\rho_s - \rho_f)/3\rho_f^2]^{1/3} \tag{1}$$

where a_1 is a function of c_1:

$$c_1 = \frac{d_p}{12[3\mu_f^2/4g\rho_f(\rho_s - \rho_f)]^{1/3}} \tag{2}$$

For c_1 less than 10, a_1 is in the range of 3-4. Above a value of $c_1 = 10$:

$$a_1 \approx 0.04c_1 + 2.6 \tag{3}$$

Soo (19) explores this question in a little more detail.

Another approach to describing the flow is to focus on particle-particle interaction. In these terms 'dilute-phase' transport has been considered to occur when the solids volume fraction is less than 0.1. (19) The transition to dense-phase flow will occur when the particles start to interact with one another. This approach is quite satisfactory for vertical transport when the gravitational force acts along the direction of flow. Under these conditions Farias and Irons (20) suggest that the key factor is whether or not particle boundary layers overlap. When the loading is low the particles and gas behave as two separate phases. As loading increases the point is reached where the boundary layers of the particles overlap, essentially leaving no 'free' gas phase. Under these conditions the gas and particles flow as one coupled phase which has important consequences for the injection regime as will be seen below. In the case of horizontal flow however the question of saltation discussed above must be considered. Farias and Irons calculate the critical loading in terms of the solids volume fraction as a function of particle Reynolds number:

$$\varepsilon_p = \frac{\pi}{6\left(\frac{6}{\sqrt{2Re_p}}+1\right)^3} \tag{4}$$

This relationship is shown as a solid line in Figure 1.

A final point concerns the relative velocity of particles and gas. A large body of evidence suggests that particle velocities can be considerably less than the gas velocity. Work by Engh et al (21), Ghosh and Lange (22) and others conclude that particle velocities are about half that of the gas. However there is some controversy over this point. (20) The measurement of particle velocities is difficult due to the problems in observing the particles themselves as well as determining whether or not sufficient length of pipe has been allowed for full development of the flow. The orientation of the pipe could also be expected to influence the measurement. Work by Klinzing and Mathur (23) has concentrated on measurement of the slip velocity $(u_f - u_p)$ as a key characterization of gas-particle flow. Their results show that the slip velocity at first increases with loading and then decreases. In the case of pulverized coal (30% minus 200 mesh) the maximum slip velocity, 5.5 m/s, occurs at a wt/wt loading of 5. This clearly fits in with the idea of coupled flow as proposed by Farias and Irons (20) and may go some way to explaining the discrepancies noted in the literature. Wen (16,17) has proposed the following equation for the calculation of particle velocity given that the value of the solid friction coefficient, λ, is known:

$$\frac{3}{4}\left(\frac{C_d}{\varepsilon^{4.7}}\right)\left(\frac{\rho_f}{\rho_p}\right)\left(\frac{(u_f-u_p)^2}{d_p g}\right)-\frac{\lambda}{2}\left(\frac{u_p^2}{D_p g}\right) = \sqrt{\frac{4}{3}\left(\frac{g d_p}{u_p^2}\right)\left(\frac{\rho_p-\rho_f}{\rho_f}\right)\left(\frac{\varepsilon^{4.7}}{C_{dt}}\right)} \tag{5}$$

Values of λ for gas-particle flow are reported in the range $2(10^{-3})$-0.1. (16,17,23)

Although this is only a very superficial review of the factors involved in particle-gas transport it does suggest that the it cannot necessarily be assumed that the flow is homogeneous. Furthermore even if it is, there are two types of homogeneous flow, one in which the gas and solids flow independently and the other in which particle boundary layers bind the flow into more of a single 'fluid'. This depends on the loading and flow orientation. All of this has a strong influence on the behaviour of the stream once it enters into the liquid.

Dynamics of Single Particle Injection

Before dealing with the 'jet'-liquid interaction it is important to review the dynamics of injection of a single particle into a liquid. This will be a valuable aid in the understanding of the forces involved in the overall injection process.

Physics of a Single Particle-Liquid Interaction

Table II summarizes the experimental conditions in a number of recent studies of single particle-liquid interaction. When a moving particle impacts on a liquid surface it will follow one of several paths depending on the circumstances:

a) if the velocity is insufficient it may be unable to penetrate into the liquid and simply remain at

the interface,
b) it may impact the surface forming a cavity and immediately spring back out into the gas,
c) it may form a cavity and pass through into the liquid on its own, or
d) form a cavity and pass through into the liquid drawing gas with it.

TABLE II. Experimental Studies of Single Particle Injection

LIQUID	PARTICLE		PARTICLE ENTRY	REFERENCE
	Composition	Diameter (mm)	VELOCITY (m/s)	
water	wood	9.5-5.08	6-8.5	26
mercury	glass stainless steel sapphire	1.6-9.5	-	25
water, glycerin	polystyrene	2, 3	3-65	27

A force or energy balance can be used to calculate the critical particle size that will penetrate a liquid surface under given conditions (9,20,24). Accounting for the inertial force, surface tension force and pressure and drag forces of the liquid on the particle Engh et al (24) derive the following equation for the critical particle radius, r_c:

$$\frac{r_c u_p^2}{8\sigma_{l/g}} = \left(1 + \frac{(\sigma_{1/3} - \sigma_{3/g})}{\sigma_{l/g}}\right)\exp\left(\frac{3}{4\rho_p}\right) + \left(1 - \frac{(\sigma_{l/3} - \sigma_{3/g})}{\sigma_{l/g}}\right) - \left(\frac{8\rho_p}{3}\right)\left(\exp\left(\frac{3}{4\rho_p}\right) - 1\right) \quad (8)$$

Farias and Irons (20) have shown that the following equation can give a better fit to the data of Ozawa et al (25) for $f_a = 1$:

$$\varepsilon_p^2 + \varepsilon_p\left(\frac{\rho_l}{\rho_p} - 1\right)\left(\frac{\mathring{\varepsilon}_p}{1 - R_E}\right) - \frac{\rho_l}{\rho_p}\left(\frac{\mathring{\varepsilon}_p}{1 - R_E}\right) = 0 \quad (9)$$

where

$$R_E = \frac{12\sigma_{l/g}(f_a - \cos\phi)}{\rho_p d_p \mathring{u}_p^2} \quad (10)$$

Gutherie et al (26) developed a mathematical model of the subsurface motion of the particle by equating the rate of change of momentum to the force balance:

$$m_p \frac{du_p}{dt} = gV_p(\rho_p - \rho_f) - m_f g - F_D - F_A - F_H \quad (11)$$

which made a reasonable prediction of particle 'trajectory', depth of penetration and residence time below the surface. When Ozawa et al (25) dropped spheres into mercury they observed that below a critical velocity the sphere would jump back out of the surface impact cavity. Above that velocity the sphere penetrated into the liquid where its behaviour is presumably described by Eq. (11). In their study Lee and Tokuda (27) observed a critical velocity below which the particle alone penetrated into the liquid. Above this velocity, approximately 13-20 m/s depending on the situation, the particle drew a columnar bubble into the liquid with it. The diameter of the bubble was 2-5 times that of the particle and the volume of the gas 10-70 times the particle volume generally increasing with particle velocity. Lee and Tokuda developed a mathematical model based on an energy balance to determine the fraction of particle kinetic energy dissipated on penetration through the surface. The energy balance is given by:

$$E = E_k + E_p + E_s + E_d + E_{pot} + E_{sr} + E_{heat} + E_{wet} + E_{sp} \qquad (12)$$

Their analysis showed that E_{pot} is negligible and E_{sr} is only 7-9% of the total input energy. In general most of the energy was found to dissipate at the surface (70-95%) mainly as heat and in the formation of a spray of liquid droplets. As a result the penetration of the particle into the liquid is only fractionally increased with increases in velocity. The authors point out however that in the case of powder injection the overall situation may be quite different due to the creation of a gas-liquid zone at the interface by gas entrainment.

Finally, it should be noted that in the case of a reactive particle the situation can change dramatically. For example, in slag fuming pulverized coal is injected into slag at 1200 °C. The coal will pyrolyze on contact with the slag releasing volatile gaseous species. Similarly, magnesium will vaporize on injection into liquid steel. In each case the tendency to release of gas on contact can influence the entrainment dynamics.

The Ternary Interfacial Energy Diagram

The ultimate fate of a particle in a gas-liquid system, after all of the inertial forces have been dissipated, is determined by interfacial energy values. In accordance with thermodynamics the system will move to minimize the total interfacial energy. Conochie and Robertson (28) have presented a general analysis of this question and developed a diagram, Figure 2, showing five possible regions as a function of the interfacial surface tensions. When X, Y or Z is greater than 0.5 then one of the surface energies is greater than the sum of the other two. In this situation the high surface energy interface will disappear in favour of the other two. Thus, when the interfacial tension between solid (phase 3) and gas is high the system will reach the lowest energy state in a 'dispersion' of separate particles and gas bubbles. If the particle-liquid interfacial energy is high then the system will move to eliminate particle-liquid contact and place the particles inside the bubble in the gas phase. When X, Y and Z are less 0.5 then the particles will sit on the liquid-gas interface. If X > Y (and therefore the liquid-solid energy is greater than the solid-gas energy) the particle will tend to lie more in the gas phase than the liquid. If Y > X then the particle will be predominantly in the liquid.

Dynamics of Submerged Powder Injection

Although the dynamics of a single particle-liquid interaction are important they do not fully describe the phenomena of a multi-particle gas jet entering a liquid. This is because the particles are not moving through a stagnant gas but instead are travelling along with a moving gas stream. This has two important consequences. Firstly, the gas flow will to some extent determine the form of gas-liquid interface which the particles will see. (29) At low gas velocities the gas will tend to enter the liquid as bubbles. At high flow rates the gas will tend towards a 'jetting' regime. Secondly, as discussed above, above a certain critical loading the solids can no longer be considered as discrete particles in the gas stream. Instead the particles and gas are intimately coupled. (20)

Lance Injection

By far the largest number of studies have been done on the injection of powders through vertical or near-vertical lances which are characteristic of most applications in steelmaking, the Mitsubishi process and Sirosmelt/Isasmelt technology. (9,20,24,30-38) Farias and Irons (20) have written a comprehensive review of the subject covering the work in the field to 1985. They developed the diagram shown in Figure 3 to delineate the injection regimes encountered in vertical downward submerged gas-particle injection. The left column gives gas velocity in terms of Mach number. The adjacent column shows the expected behaviour of a gas-only stream as velocity is increased. Jetting (as indicated by the cut off cone) only appears at supersonic velocity for higher gas-to-liquid density ratios. Fine particles will be closely coupled to the gas phase in normal injection situations. As a result the flow tends to behave as a single phase with a relatively high momentum leading to jetting at relatively low Mach numbers. The boundary between bubbling and jetting as a function of particle Reynolds number and volume fraction is shown in Figure 1. For coarse particles the gas and particles are not necessarily coupled and the gas tends to behave independently following the pattern of gas only injection. The particles will generally penetrate through the interface if they are wetted by the liquid. Once in the liquid the particles will either move as a 'particle-liquid' jet or as individual settling particles if particle-liquid coupling is poor. Farias and Irons also present a model for submerged vertical downward jets which accounts for the effect of buoyancy on the flow and explains the small cone angles usually observed in this situation. The value of the entrainment constant in the model was set at about 20% of that for single phase gas jets or high Reynolds number horizontal gas jets in liquid.

A slightly different situations exists for unsubmerged lances. (39,40) In this case at low gas velocities a surface cavity forms instead of a bubble. Kimura (39) reports a criteria for cavity formation based on the parameter N_j:

$$N_j = N_p \mathring{u}_g A_p \left(\frac{\rho_p}{\rho_l} \right) \tag{13}$$

Below $N_j = 1000$ the particle jet simply forms a cavity on the surface. Above a value of 2000 the jet penetrates into the liquid carrying both particles and bubbles below the surface. More recently Fukunaka et al (40) have examined the same question focussing on the extent of gas entrainment and jet penetration under these conditions. They show that for boundary layer thicknesses of up to 10 particle diameters the all of the gas essentially follows the particles into the liquid.

Finally it is interesting to note that, as in the case of single particles, powders injected at low velocities will not necessarily penetrate the interface. (20,22)

Horizontal and Vertical-Upward Injection

In contrast to the work on lance injection there have been relatively few studies concerned with the horizontal and vertical-upward injection of particle-gas mixture. Engh and Bertheussen (30) calculated the trajectory of a particle-gas jet in a melt using the model of Themelis et al (39) of a jet entraining liquid and expanding at a constant cone angle. For horizontal gas-particle jets Engh et al (21) found that the use of cone angles in the range 2.5-7° fit the measured penetration in water modelling experiments relatively well.

A few studies have examined the influence of solids on the vertical-upward injection process. (41-43) In general the conclusions are as discussed for lance injection. Mori and Ozawa (43) conclude that the transition from bubbling to jetting is reduced to lower Mach numbers in gas-particle injection. The presence of the solids tends to reduce the cone angle of the jet in the liquid and increase the measured bubble frequency. It is worthwhile to note that the buoyancy force and the stream momentum act in the same direction. As a result during injection in the bubbling regime the gas-particle stream will produce an 'inverted cavity' in the top of the bubble.

Heat and Mass Transfer Phenomena

Although the major concern of this review has been a discussion of the dynamics of gas-particle injection it is worth noting that the entrainment of powders in the liquid bath is only one aspect of the overall process. During the injection process the particles and gas will be heating up due to heat transfer from the liquid. If the gas-particle flow is uncoupled and the gas is in a bubbling regime then the powder will pass through a bubble once it leaves the lance or tuyere tip. It will undergo heating and if the gas is potentially reactive then there is the chance that a reaction can start during this period. For example, in slag fuming coal is injected with air into molten slag and combustion of the mixture might be expected to start before the coal can entrain in the slag. The reduction efficiency of the coal would be expected to fall as a result. Once the powder is entrained in the bath it will heat up as a function of heat transfer from the surrounding fluid and at the same time react with the bath through some mechanism likely involving liquid phase mass transfer. In most cases the injected solid is lighter than the liquid and will rise to the surface of the melt. At this point it may become part of a floating slag layer, as in the case of lime injection into steel, or leave the system, as in coal injection for slag reduction. In each case the efficiency with which the solid is used is in part a function of its residence time in the melt. Not surprisingly this in turn is related to the injection conditions themselves. For example, as discussed above, in lance injection the penetration of the gas-particle jet is a function of jet momentum and gas-particle coupling. If the particles are not coupled to the gas then they will tend to enter the liquid alone. This may be desirable if they will react with the carrier gas at high temperatures.

It should also be noted that in the lance injection of powders there will be a certain degree of heat transfer from the bath through the walls of the lance to the flowing gas-particle mixture. It has been shown that the rate of heat transfer to the mixture can be significantly different than to a gas alone. (36,44,45) The particles increase the specific heat capacity of the flow as well as influence the rate of heat transfer from the walls.

Another important aspect of the heat and mass transfer phenomena is the question of accretion formation on the injector, whether it be a lance or tuyere. The formation of accretions is a critical phenomena in injection processes because a stable, non-blocking accretion can be used to prolong refractory life as in the shrouded tuyere for oxygen steelmaking or lance life as in SIROSMELT technology. In other processes, such as the copper converting, accretions will often block the gas flow and require punching. (46) Several papers have commented on the question of clogging during lance injection and its relationship to bubbling and jetting, and heat transfer. (33,38,45)

Summary and Conclusions

The fundamental aspects of gas-particle injection processes involve four inter-related areas:
a) gas-particle flow in the injector,
b) the physics of single particle-liquid interaction,
c) the dynamics of powder-liquid interaction, and
d) heat and mass transfer phenomena.

If the gas and particles are flowing in a closely coupled manner they will tend to behave as a 'single phase' jet on entry into the liquid. The presence of the particles will also cause jetting rather than bubbling to occur at lower injection velocities. In general, the gas will be largely entrained with the particles in these circumstances. However, if the flow is uncoupled, then the particles will behave independently and their entrainment in the liquid will depend more on the dynamics of single particle-liquid interaction. The slip velocity will therefore be a more important parameter. It is also worth noting that there are differences in vertical and horizontal gas-particle flow. In horizontal flow non-homogeneous conditions can develop at high loadings if the gas velocity is not kept above the saltation velocity. The presence of a dense, slower moving suspension on the bottom of the pipe could dramatically reduce particle entrainment in the liquid. Finally, heat and mass transfer phenomena cannot be ignored. The injection of a relatively cold gas-particle mixture into the liquid results in local cooling and the potential formation of accretions which may block the injector.

Nomenclature

a_1	variable calculated in Eq. (3)
A_p	projected area of one particle
c_1	variable calculated in Eq. (2)
C_d	drag coefficient of single particle evaluated at slip velocity
C_{dt}	drag coefficient of single particle evaluated at terminal free fall velocity
d_p	particle diameter
D	pipe diameter (in inches)
D_p	pipe diameter
E	energy of particle before impact with surface
E_d	energy dissipated in liquid flow
E_{heat}	heat energy generated by particle impact
E_k	kinetic energy of particle in the liquid
E_p	potential energy of the columnar bubble
E_{pot}	gravitational potential energy of the cylindrical cavity wave
E_s	surface energy of the columnar bubble
E_{sp}	energy of spray of droplets released by particle impact
E_{sr}	surface energy of the cylindrical cavity wave

E_{wet}	net energy in wetting particle surface
f_a	ratio of extra surface area created at the gas-liquid interface as the particle passes through to the area of the particle
F_A	drag force from 'added mass'
F_D	drag force from steady translation
F_H	drag force from liquid history term
g	gravitational acceleration
m_f	mass of fluid displaced by particle
m_p	mass of particle
Ma	Mach number
N_j	cavity injection number
N_p	number of particles per unit volume of gas
R_E	energy ratio defined by Eq. (10)
Re_p	particle Reynolds number
u	velocity
u_s	saltation velocity
V_p	volume of particle
X	$\sigma_{l/3}/\Sigma\sigma$
Y	$\sigma_{3/g}/\Sigma\sigma$
Z	$\sigma_{l/g}/\Sigma\sigma$
ε	gas volume fraction
ε_p	particle volume fraction in gas stream
λ	solid friction coefficient for particle flow
μ	viscosity
φ	liquid-third phase (particle) contact angle
ρ	density
$\sigma_{l/g}$	liquid-gas interfacial energy
$\sigma_{3/g}$	third phase (particle)-gas interfacial energy
$\sigma_{l/3}$	liquid-third phase (particle) interfacial energy
θ	third phase (particle)-gas contact angle
θ_p	particle volume fraction (Fig. 1)
$\Sigma\sigma$	$(\sigma_{l/g} + \sigma_{3/g} + \sigma_{l/3})$

Subscripts

f	fluid
p	particle
s	solid phase or particle

Superscripts

$^{\circ}$	value at nozzle tip

References

1. E.G. Schempp, F. Eichinger, J.R. Grossmann and E. Fritz, Iron and Steel Engineer, August, p. 30, (1986).
2. Y. Nakajima, Y. Fukami, T. Moriya and S. Maruhashi, in 72nd Steelmaking Conference Proceedings, ISS-AIME, p. 469 (1989).
3. E.T.R. Jones and J.K. Batham, in 72nd Steelmaking Conference Proceedings, ISS-AIME, p. 537 (1989).
4. A. Ishizaka, O. Terada, A. Yamagami and T. Hasegawa, in 72nd Steelmaking Conference Proceedings, ISS-AIME, p. 249 (1989).
5. H. Okuda, M. Suitoh, H. Take and F. Sudo, in 71st Steelmaking Conference Proceedings, ISS-AIME, p. 249 (1988).
6. H. Klein, J.F. Liesh, H. Iso and K. Nakamura, in 68th Steelmaking Conference Proceedings, ISS-AIME, p. 129 (1985).
7. Anonymous, Steel Times, Feb., p. 83 (1986).
8. M. Ozawa, S. Kitagawa, S. Nakayama and Y. Takesono, Trans. I.S.I.J., vol. 26, p. 621 (1986).

9. R.J. O'Malley, C.E. Dremann and D. Apelian, J. Metals, p. 14 (1979).
10. Y. Mori and T. Kimura, Metall. Review M.M.I.J., vol. 3, p. 141 (1986).
11. Y. Okajima, T. Kimura, Y.Mori and T. Kusakabe, Metall. Review M.M.I.J., vol. 3, p. 90 (1986).
12. S.P. Matthew, G.R. McKean, R.L. Player and K.E. Ramus, in Lead-Zinc'90, T.S. Mackey and R.D. Prengaman, ed., AIME, p. 889 (1990).
13. J.M. Floyd, K.W.Jones, W.T. Denholm, R.N. Taylor, R.A. McCelland and J. O'Shea, in Symposium on Extractive Metallurgy, Australas. I.M.M., p. 25 (1984).
14. G.G. Richards and J.K. Brimacombe and G.W. Toop, Metall. Trans. B, vol. 16B, p. 513 (1985).
15. R. Bassa, A. del Campo and C. Barria, in Copper 87, vol. 4, p. 149 (1987).
16. C.Y. Wen, in Bulk Materials Handling, Vol.1, M.C. Hawk, ed., p. 258, University of Pittsburgh, Pittsburgh (1971).
17. C.Y. Wen and W.S. O'Brien in Gas-Solids Handling in the Process Industries, J.M. Marchello and A. Gomezplata, ed., p.89, Marcel Dekker, New York (1976).
18. F.A. Zenz, Ind. Eng. Chem. vol. 3, p. 65 (1964).
19. S.L. Soo, in Handbook of Multiphase Systems, G. Hetsroni, ed., p. 7-3, McGraw-Hill, New York (1982).
20. L.R. Farias and G.A. Irons, Metall. Trans. B, vol. 16B, p. 211 (1985).
21. T.A. Engh, K. Larsen and K. Venas, Ironmaking and Steelmaking, vol. 6, p. 268 (1979).
22. D.N. Ghosh and K.W. Lange, Ironmaking and Steelmaking, vol. 9, p. 136 (1982).
23. G. Klinzing and M.P. Mathur, in Fluidization and Fluid Particle Systems: Theories and Applications, T.M. Knowlton, ed., p. 24, Amer. Inst. Chem. Eng. (1984).
24. T.A. Engh, H. Sandberg, A. Hultkvist and L.G. Norberg, Scan. J. Met., vol. 1, p. 103 (1972).
25. Y. Ozawa, K. Suzuki and K. Mori, Tetsu-to-Hagane, vol. 69, p. 753 (1983).
26. R.I.L. Gutherie, R. Clift and H. Henein, Metall. Trans. B, vol. 6B, p. 321 (1975).
27. J-G. Lee and M. Tokuda, Trans. I.S.I.J., vol. 28, p. 278 (1988).
28. D.S. Conochie and D.G.C. Robertson, Trans. Instn. Min. Metall., vol. C89, p. C61, (1980).
29. E.O. Hoefele and J.K. Brimacombe, Metall. Trans. B, vol. 10B, p. 631 (1979).
30. T.A. Engh and H. Bertheussen, Scan. J. Met., vol. 4, p. 241 (1975).
31. G. Carlsson, in Injection Phenomena in Extraction and Refining, A.E. Wraith, ed., University of Newcastle upon Tyne, p. C1 (1982).
32. M.J. McNallan, J.O. Park and Y.W. Chang, in Scaninject III, MEFOS, Lulea, p. 9:1 (1983).
33. G.A. Irons and B-H. Tu, in Scaninject III, MEFOS, Lulea, p. 11:1 (1983).
34. B-D. You, W. Polanschutz and H. Hiebler, in Scaninject III, MEFOS, Lulea, p. 10:1 (1983).
35. Y.W. Chang, J.O. Park and M.J. McNallan, in 4th Process Technology Conference, ISS-AIME, p. 23 (1984).
36. G.A. Irons, in Scaninject IV, MEFOS, Lulea, p. 3:1 (1986).
37. L.R. Farias and G.A. Irons, Metall. Trans. B, vol. 17B, p. 77 (1986).
38. Y. Shirrota, K. Marukawa, N. Hiroki and H. Ikemiya, Trans I.S.I.J., vol. 27, p. 18, (1987).
39. N.J. Themelis, P. Tarassoff and J. Szekely, Trans. A.I.M.E., vol. 245, p. 2425 (1969).
40. E. Kimura, Trans. I.S.I.J., vol. 23, p. 523 (1983).
41. Y. Fukunaka, T. Shimada, T. Nishihara, H. Tanemura and Y. Kondo, Metall. Trans. B, vol. 20B, p. 499 (1989).
42. A.H. Castillejos, M.Sc. (Eng.) Thesis, Imperial College of Science and Technology, London (1979).
42. D.G.C. Robertson, D.S. Conohie and A.H. Castillejos, in Scaninject II, MEFOS and Jernkontoret, Lulea, p. 4:1 (1980).
43. K. Mori and Y. Ozawa, in Injection Phenomena in Extraction and Refining, A.E. Wraith, ed., University of Newcastle upon Tyne, p. I1 (1982).
44. C.R. Depew and T.J. Kramer, in Advances in Heat Transfer, vol. 9, T.F. Irvine, Jr. and J.P. Hartnett, eds., p. 113 (1973).
45. G.A. Irons, Metall. Trans. B, vol. 18B, p. 105 (1987).
46. A.A. Bustos, J.K. Brimacombe and G.G. Richards, Canadian Metallurgical Quarterly, vol.27, p. 7 (1988).

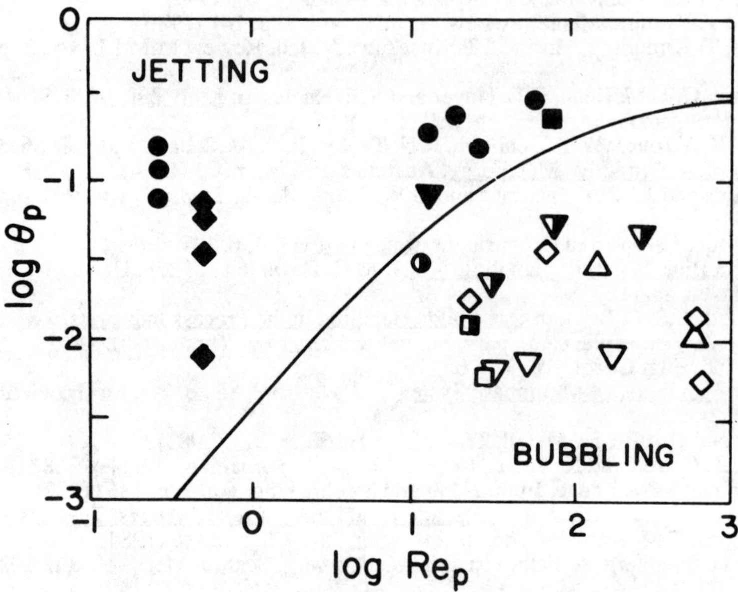

Figure 1. Jetting-bubbling diagram for gas-particle flow (from reference 20).

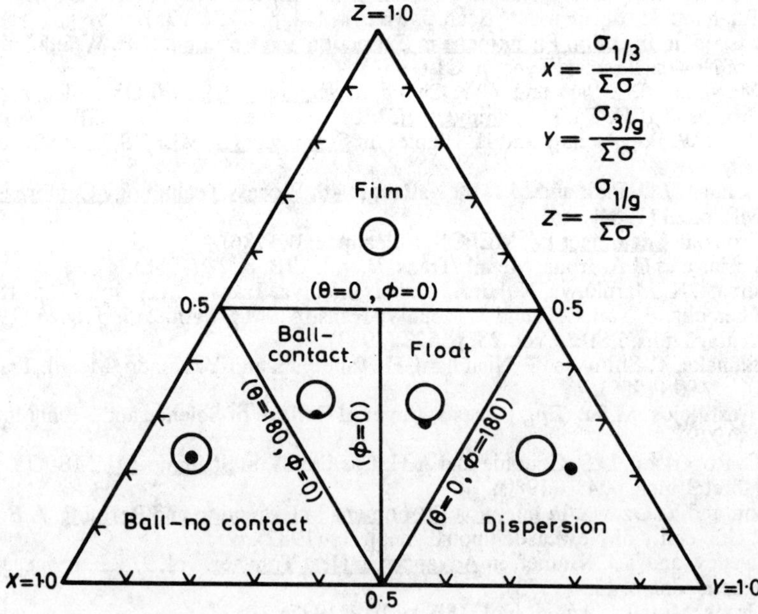

Figure 2. Ternary interfacial energy diagram for gas-liquid-third phase (from reference 29)

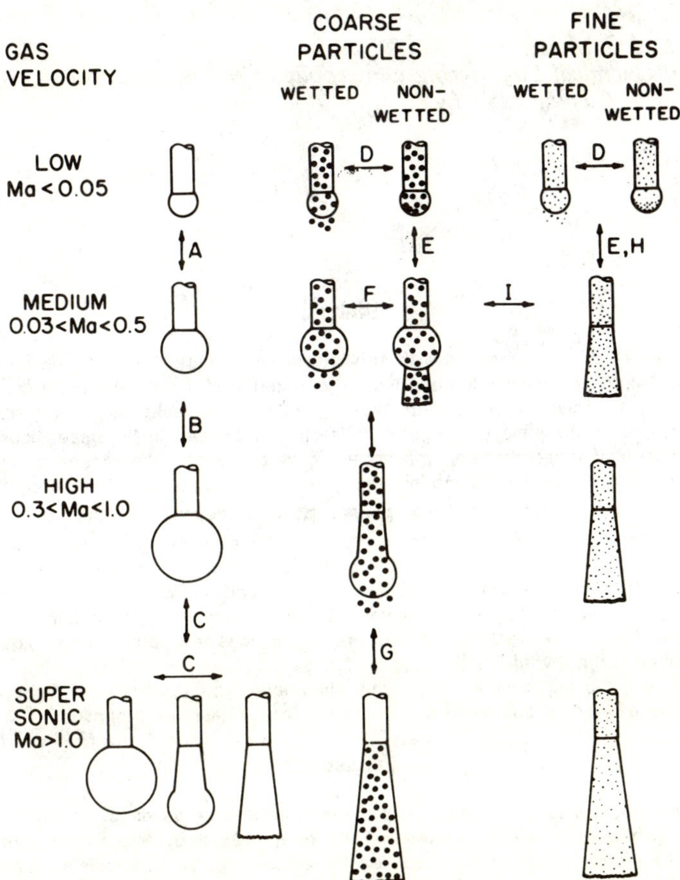

Figure 3. Injection regimes for gas-particle submerged lance injection (from reference 20).

The behaviour of large gas bubbles at a liquid-liquid interface. Part-1: the entrainment of liquid drops

R. Kondur and J.M. Shaw
Department of Chemical Engineering and Applied Chemistry, University of Toronto
Toronto, Ontario, Canada M5S 1A4

Abstract

Gas agitated liquid-liquid dispersions arise in applications as diverse as processes for the manufacture of iron and steel and for the hydrogenation of bitumen and coal. In the present work, some consequences of gas addition to slag/matte type systems are investigated. The size distribution of liquid drops entrained by the passage of gas bubbles from the lower to the upper liquid phase and the subsequent breakage of these drops in the upper liquid phase are phenomena of particular interest. Experiments were conducted with the model ternary system: decane (50 wt %) + sunflower oil - water (67 wt %) + sucrose -air. The two liquid phases possess densities of 806 and 1200 Kg/m^3 and viscosities of 5.2 and 6.6 x 10^{-3} Kg/ms respectively. Apart from interfacial tension, this system is a suitable organic analogue for modelling the hydrodynamic behaviour of copper converters, for example, where the density difference between the two liquid phases and the viscosities of the two liquid phases are comparable. Reynolds similarity is achieved. Dynamic drop size distributions in the upper liquid phase, determined through the analysis of video and still images, are found to be in general agreement with the predictions of a hydrodynamic model for gas agitated liquid-liquid dispersions. General guidelines concerning the introduction of gas into slag/matte type systems and the operation of batch vessels that minimize net drop entrainment in the upper liquid phase are proposed.

Introduction

Gas agitated liquid-liquid dispersions arise in processes as diverse as direct coal liquefaction and coal oil coprocessing (Shaw *et al.* 1988b, Dukhedin-Lalla *et al.*, 1989), pyrometallurgical processes for the production of nickel and copper, and processes for electro-organic synthesis. However, the steady state and dynamic behaviour of such dispersions has received comparatively little attention in the open literature where the focus has been the behaviour of vertical tubular vessels for a limited range of fluid properties (Yoshida and Yamada, 1972; Hatate *et al.*, 1976; Kato *et al.*, 1984) and consequently, the behaviour of these dispersions is poorly characterized. More recently (Shaw *et al.*,1988; Hatzikiriakos *et al.*, 1990a, Hatzikiriakos *et al.*, 1990b) showed that the analogies frequently employed between the steady state and dynamic behaviour of mechanically agitated and gas agitated dispersions are inappropriate. The authors also introduced semi empirical correlations for predicting the maximum and minimum drop sizes stable against break-up and coalescence and steady state Sauter mean diameter as functions of operating variables and the physical properties of the fluids.

The transitional behaviour of dispersed phase drops in liquid-liquid dispersions is also of great interest, as in many applications mean residence times are too short for steady state drop size distributions to obtain (Lee *et al.*, 1984). The transition arising as the dispersed phase first enters a vessel and is broken up is of particular importance. The prediction of dynamic break-up and coalescence rates for dispersed phase drops in mechanically agitated vessels has been addressed previously (Shah and Ramkrishna, 1973; Park and Blair, 1975; Coulaloglou and Tavlarides, 1976; Coulaloglou and Tavlarides, 1977; Narsimhan *et al.*, 1979; Stamatoudis and Tavlarides, 1985). These workers approached the problem of finding proper forms for the break-up and coalescence rate functions, using population balances. Mean drop sizes and drop size distributions are readily predicted as functions of time, if these functions are known. The principal obstacle in this approach is the expression of break-up and coalescence rates in terms of the hydrodynamics and physical properties of a particular mechanically agitated system. An alternative approach is to determine the transition probability function characterizing drop breakage. Ramkrishna (1973) employed a similarity concept to define the transition probability function and found that experimental transitional cumulative drop size distributions could yield useful information for predicting the transient interfacial area in mechanically agitated liquid-liquid dispersions. He tested the similarity transformation between the drop size variable and time and found that it fit the experimental data of Madden and McCoy (1969) reasonably well. Narsimhan (1980) generalized the concept of similarity by expressing the cumulative size distribution function of daughter drops in terms of a transition probability function. By so doing, he assumed that larger drops break up into more daughter drops than smaller ones. The latter were assumed to break up into two equal size daughter drops. More recently, the similarity concept has been applied to the breakage of dispersed liquid phase drops in lean, gas-agitated dispersions (Hatzikiriakos *et al.*, 1990b). In this latter case the impact of drop coalescence was neglected, and no a priori assumptions were made concerning probability of drop breakage or the nature of the breakage function. Despite these restrictions mean drop sizes and drop size distributions for diverse organic liquid-liquid systems were well represented.

In this paper, the mathematical model devised for batch gas-agitated liquid-liquid dispersions (Hatzikiriakos *et al.*, 1990b) is extended to include the evolution of liquid drop size distributions and the concentration of dispersed liquid drops in the upper liquid phase of slag-matte type systems. The impact of drop coalescence per se is neglected, as the evolution of drop size distributions, at short times, is predominantly a function of drop entrainment by gas bubbles, drop breakage in the slag phase and the residence time of the drops in the slag phase. The model provides an upper bound for the extent of entrainment.

The Prediction of Dynamic Drop Size Distributions

1. The Role of Similarity in Drop Breakage

Dynamic drop size distributions can be estimated from population balances. The population balance equations may be written in terms of drop break-up and coalescence rates or in terms of probability functions. The former approach is more difficult to apply as noted above. The probability functions of interest in the present case include the transition probability function of drop break-up and the size distribution of drops originating from the break-up of a parent drop. The transition probability function of drop coalescence is neglected.

The change in the cumulative drop size distribution with respect to time for drops entrained by a single bubble can be expressed as

$$\partial F/\partial t = \int_V^\infty \Gamma(V')G(V,V')dF(V',t) \tag{1}$$

where: the volume fraction of drops with a volume less than V at time t is denoted as F(V,t) a cumulative distribution function that satisfies the conditions:

$F(V,t)=0$ when $V \leq 0$; $F(V,t)=1$ when $V \to \infty$

$\Gamma(V')dt$ represents the probability that a drop of volume V' breaks in the time interval (t,t+dt). If a parent drop of volume V' breaks up into smaller drops then the volume of these drops with volume less than V is represented by a cumulative distribution function G(V,V'), that satisfies the conditions:

$G(V,V') = 0$ when $V \leq 0$; $G(V,V') = 1$ when $V \geq V'$

as long as all of the drops remain entrained in the upper liquid phase.

Both functions $\Gamma(V')$ and $G(V,V')$ must be defined in order to evaluate how cumulative drop size distributions change with respect to time. Since cumulative drop size distributions are known experimentally, one may assume appropriate functional forms for $\Gamma(V')$ and $G(V,V')$ and evaluate the constants appearing in them from experimental data. Conversely, the functional forms of $\Gamma(V')$ and $G(V,V')$ can be inferred from experimental distributions.

Ramkrishna (1973) assumed that the break-up probability function followed a simple power law of the form:

$$\Gamma(V) = KV^n \tag{2}$$

where K and n are coefficients to be identified from experimental measurements. A power law is consistent with much of the data available and given the uncertainty of the data, there is little justification for a more sophisticated model. K, for example, can be expressed as a function of experimental conditions (Das et al., 1987). Few restrictions need be imposed on the functional form of G(V,V') (Narsimhan et al., 1979) although similar breakage, which implies that both large and small drops break-up similarly,

$$G(V,V') = g(V/V'); \quad 0 < V/V' < 1 \tag{3}$$

has been assumed for simplicity. With these assumptions, Equation 1 can be recaste as:

$$\partial F/\partial t = \int_V^{\infty} KV'^n G(V,V')dF(V',t) \tag{4}$$

A similarity variable $Z=(1+Kt)V^n$ exists for Equation 4 (Fillipov, 1961). For times sufficiently large this variable may be approximated as KtV^n, which implies that for a fixed value of F(V,t) the similarity variable $Z=KtV^n$ remains constant.

Shaw *et al.*, 1990b showed that for batch gas-agitated liquid-liquid dispersions the appropriate similarity variable is

$$Z = K' \; t \; [V / V_{vs}] \; [\mu_d/\mu_c]^a \tag{5}$$

and that transitional drop size distributions are well characterized as a Weibull distribution in Z:

$$F(Z) = 1 - \exp\{ -A [1/3 \ln Z]^C\} \tag{6}$$

Optimum coefficient values for diverse experimental dispersions, determined using a nonlinear optimization routine, were 0.2936, 1.85×10^{-7}, 10.55, and 135 s^{-1} for a, A, C, K' respectively. The transition period pertains when $V_{vs,\,t}/V_{vs} > 1$ or a comparable criterion is met.

2. Estimation of mean drop size at steady state

The applicability of equation 6 is predicated on the availability of values for mean drop size at steady state as a function of operating conditions and vessel geometry. Such values can be obtained from a fundamental evaluation of the prevailing shear fields. In order for drops to become unstable in a shear field, we have from Hinze (1955)

$$\tau > \sigma/d + \{\mu_d/d\}\{\tau/\rho_c\}^{1/2} \tag{7}$$

where $\tau \sim$ force/area $\sim$ energy/volume

On a microscopic level we can consider a single bubble-drop interaction, with a prevailing energy dissipation rate, ε, in the continuous fluid in the region of the rising bubble, and with a characteristic duration, t'. Only a fraction of the available energy can be dissipated by the dispersed phase through internal circulation, or through the creation of additional interfacial area. Much of the energy has an inappropriate length scale and merely causes drops to convect. Thus, we may write

$$\varepsilon_{eff} \, \rho_c \, t' > \sigma/d_{vs} + \mu_d/d_{vs} \{ \varepsilon_{eff} \, t \}^{1/2} \tag{8}$$

The effective energy per unit volume, $\varepsilon_{eff} \rho_c t'$, can be expressed as a function of the energy per unit volume of fluid per collision , the probability that a bubble will cause a drop to rupture, and the energy fraction available for drop rupture. Each of these terms can be evaluated by applying fundamental fluid mechanic concepts to the behaviour of individual bubbles and drops (Hatzikiriakos *et al.*, 1990a) yielding an implicit expression for mean drop size at infinite dilution:

$$\tau = 1.5\pi\rho_c g D \underbrace{\{8\pi^2 c U_g H\sigma\sigma'/D\mu U_b^2\}}_{\leq 1} \underbrace{\{d_{vs}/T\}^6}_{\leq 1} > \sigma/d_{vs} + \{\mu_d/d_{vs}\}\{\tau/\rho_c\}^{1/2} \tag{9}$$

The impact of drop-drop coalescence on mean drop size at steady state can be estimated from the probability of binary, ternary and quaternary collisions, which in turn are functions of dispersed phase concentration. Higher order terms may be neglected. As slip velocities for drops are low, drops must be in close physical contact before coalescence can occur. If each drop is assigned a cubic cell in a dispersion, interactions with 6 face neighbours, and 12 edge neighbours must be considered. The probability that two drops are within a distance L' can be approximated as a sum of joint probabilities, P_2 for the two types of interactions:

$$\underbrace{P_2}_{\leq 1} \approx 6 \, d_{vs} [L' + d_{vs}]/[L + d_{vs}]^2 + 3 \, d_{vs}^{\,2} [L' + d_{vs}]^2/[L + d_{vs}]^4 \tag{10}$$

where $L + d_{vs} = d_{vs} [\pi / 6 \phi]^{1/3}$ is the dimension of the cell

Similarly, the probability of a ternary interaction within a fixed distance can be expressed as:

$$P_3 \sim 2.25\, d_{vs}{}^2\, [L' + d_{vs}]^4/[L + d_{vs}]^6 \leq 1 \quad (11)$$

and a quaternary interaction as:

$$P_4 \sim 0.1875\, d_{vs}{}^2\, [L' + d_{vs}]^6/[L + d_{vs}]^8 \leq 1 \quad (12)$$

Mean drop diameter becomes a function of mean diameter at infinite dilution, ie. d_{vs} from equation 16, and the incremental probability terms:

$$d_{vs\phi} \sim d_{vs}\, [\, 1 + P_2(^3\sqrt{2} - 1) + P_3\, (^3\sqrt{3} - P_2{}^3\sqrt{2}) + P_4\, (^3\sqrt{4} - P_3{}^3\sqrt{3})] \quad (13)$$

where the probabilities are evaluated in the limit $L' \rightarrow 0$, and each term in the sum is the product of the probability of the Interaction type and the resultant increment in drop diameter. The sauter mean volume at steady state becomes:

$$V_{vs} = [\Pi/6][d_{vs\phi}]^3 \quad (14)$$

3. The Proposed Model

The proposed model for the over all drop size distribution, f(V,t), can be expressed as the sum of the drop size distributions associated with each bubble:

$$f(V,t) = \sum_{1}^{j(t)} X_i\ (t)\ F(V, V_{vs\phi}, t_i^*) / \sum_{1}^{j(t)} X_i\ (t) \quad (15)$$

and the concentration of the dispersed phase in the upper liquid phase, Φ, can be expressed as:

$$\Phi(t) = 1/[\ 1 + Y/\{v \sum_{1}^{j(t)} X_i\ (t)\}] \quad (16)$$

where X_i (t) is the fraction of the liquid entering with bubble i remaining in the upper phase at time t
t_i^* is the apparent age of the drop distribution arising from the liquid entrained with bubble i
Y is the initial volume of the upper liquid phase
j is the number of bubbles that have entered the upper liquid phase

The drop size distributions associated with bubbles age as a result of drop breakup and the elimination of larger drops from the dispersion which pass back through the liquid-liquid interface. The residence time of drops in the dispersion is determined from the slip velocity and the depth of the upper liquid phase. Individual distributions age more rapidly than indicated by equation 6 and apparent as opposed to actual time is the appropriate variable. The fraction of the liquid entrained with a bubble and retained in the dispersion decreases rapidly with time. This effect is included in the calculation of overall drop size distribution and concentration. The drop size distributions arising from each bubble, concentration, and mean drop diameter are updated at each time step. Inputs for the model include physical property data, gas flux, mean bubble diameter, the volume of liquid entrained per bubble and the maximum size of entrained drops. As with many fluid mechanic models, the algorithm is computationally intensive. The computations for this paper required 10 cpu hours on a CRAY XMP super computer.

Experimental

Experiments were conducted in a vessel with a rectangular cross-section of 0.15 X 0.30 m. The vessel, 1.00 m in height, was constructed of 6.35 mm thick sheets of perspex. Five holes were drilled in the bottom plate of the column and fitted with interchangeable nozzles (5. mm diameter). Drop and bubble diameters were determined using high speed video (Panasonic S-VHS, shutter speed .005 s) at low gas fluxes and photographically using a Nikon camera (shutter speed 0.004 s) in conjuction with an image analyzer (Bausch and Lomb, Omnicon-3000 with NOVA-4 data general) at high gas fluxes. The vessel was operated in batch mode with respect to both liquid phases. Concentrations were determined by centrifuging 20 ml samples extracted from ports along the side of the reactor.

Experiments were conducted with the model ternary system decane (50 wt %) + sunflower oil - water (67 wt %) + sucrose -air. The two liquid phases possess densities of 806 and 1200 Kg/m^3 and viscosities of 5.2 and 6.6 x 10^{-3} Kg/ms respectively. The liquid-liquid interfacial tension and the surface tension of the decane solution are 0.023 and 0.025 Kg/s^2 respectively. Viscosity was measured using an Ostwald viscometer and interfacial tension using a ring intertensiometer. During a typical experiment, the column was filled with sugar solution to a depth of 0.25 m. The organic liquid layer thickness was 0.105 m and the air flow rate 1x10-4 m^3/s. At low gas fluxes, all data were obtained while the column was operating. At high gas fluxes, the following procedure was used to determine drop size distributions. After stopping the air supply, photographs were taken immediately and the experiment was repeated, from the beginning, in order to obtain photographs for the next time interval. A typical image analysis for each experiment was based on the size analysis of 1000 drops. Size analysis of approximately 500 drops gave similar results. Mean drop diameters for duplicate experiments were repeatable to within ± 15 μm (Shaw *et al.*, 1988a). All concentration measurements were obtained while the column was operating.

Results and Discussion

The mechanisms for drop entrainment at a liquid-liquid interface, the amount of liquid entrained per bubble, and the size of the drops entrained vary with bubble size, geometry, and flux (Kondur and Shaw, 1990). For the purpose of the present work, spherical cap bubbles with a mean volume of $1x10^{-6}$ m^3 were employed. In this case, short strings of lower phase material are dragged through the plane of the liquid-liquid interface with the rising bubbles. The strings break within 0.5 s and some of the material is returned immediately to the lower phase. Much of the remainder is retained by the bubbles and then released at the upper-liquid-phase air interface. The total amount of liquid entrained per bubble is approximately $0.1x10^{-6}$ m^3 for 1 $x10^{-6}$ m^3 bubbles and the maximum drop size is 0.005 m. The general structure of the model predictions was prepared with these input data and the results are presented in Figures 1, 2, 3, and 4. Steady state is reached more rapidly when the upper liquid phase is shallow and the steady state concentration is lower than when the upper liquid phase is deep. Emulsions, which are defined arbitrarily as dispersions with a dispersed phase volume fraction exceeding 0.3, tend to form even at low gas fluxes if the upper liquid phase is deep and the operating time long enough. Mean drop diameter exhibits a complex dependence on gas flux, slag depth but becomes smaller with time until a steady state value is reached.

The tendency for the dispersions to become emulsions at steady state is clearly shown in Figure 1. Deep dispersions are more susceptible than shallow ones at fixed gas flux because of the greater probability of drop breakage. However, the time required for deep dispersions to approach steady state can exceed 1000 s and they can therefore be metastable at 1000 s as shown in Figure 5. After 2000 s, for example, a 1.6 m deep dispersion at a gas flux of 0.01 m/s becomes an emulsion, whereas the others remain stable. In the model, the bubble flux is presumed to be uniform. This cannot be achieved experimentally and the experimental fluxes correspond to the flux of the bubble plume at the liquid-liquid interface. The experimental data shown yielded a dispersion and an emulsion respectively, and illustrate the sensitivity of emulsion/dispersion formation to the value of the gas flux. A more detailed comparison between model predictions and experimental data is given in Figure 6.

The fluctuation of mean drop size with upper liquid phase depth, Figure 3, does not lead to the the prediction of an optimum upper liquid phase depth. The model predicts that the total volume of lower liquid phase entrained is minimized at low upper liquid phase depths as shown in Figure 7. The model also predicts that the time required to settle entrained material back into the lower liquid phase once bubbling has ceased is strongly dependent on upper liquid phase depth. For the example cited in Figure 8, it takes 20 times longer to recover 98 % of the entrained lower liquid phase at a depth of 0.8 m compared with a depth of 0.4 m. These results do not apply if an emulsion is formed as the settling times are much longer in this latter case.

Substitution of approximate metallurgical properties pertaining to copper converters (ρ_{slag}=3500 Kg/m^3, ρ_{matte} =4400 Kg/m^3, μ_{slag}= .003 Kg/ms, μ_{matte}=.004 Kg/ms, $\sigma_{slag\text{-}air}$ = 1 Kg/s^2, $\sigma_{slag\text{-}matte}$ = 0.1 Kg/s^2) into the model yields results with similar trends. However, the drops are smaller (mean drop sizes are in the order of 200 to 800 μm), it takes longer to reach steady state, and the dispersions have a greater tendency to form emulsions at low slag depths, ie. for a gas flux of .01 m/s and a slag height of .4 m, an emulsion is formed within 300 s, whereas for a slag depth of 1.6 m the dispersion remains metastable at 1000 s. As the physical properties of the slag and matte, and the height of the dispersion change continuously with time during a blow it is not possible to provide precise estimates for dispersion properties at the end of a blow. Consequently, the proposed guidelines are of a general nature. In order to minimize the net entrainment of matte in the slag and facilitate the separation of matte and slag at the end of a blow:

1. minimize the slag depth
2. minimize blow duration
3. reduce the maximum gas flux (ie. reduce the maximum energy dissipation rate in the vessel)
   by introducing the gas as far as possible from the liquid-liquid interface
   by eliminating chanelling
4. reduce the mean bubble size (ie. reduce the maximum shear rate in the vessel)
   this reduces the rate of drop entrainment and the rate of breakage of entrained drops

Clearly, these hydrodynamic guidelines conflict with kinetic requirements and an optimum solution must be found for individual processes.

Conclusions

A mathematical model, based on the behaviour of individual bubbles and drops has been developed to describe the impact of gas addition on the entrainment of matte into the slag phase of batch slag/matte type systems. Both dynamic and steady state conditions were modelled. The model is computationally intensive but agrees well with available experimental data. Metallurgical vessels such as copper converters are well characterized as dynamic processes from a hydrodynamic point of view as entrained dispersions tend to approach steady state conditions slowly. Consequently, four general options including reducing blow times can be exploited in order to minimize net matte entrainment in the slag phase.

Nomenclature

A, a	dimensionless coefficients
C, c	dimensionless coefficient, dimensional coefficient (c = 7.0×10^{-4} s^2 Kg^{-1})
D,d	bubble diameter, drop diameter, m
d_{vs}, $d_{vs\phi}$	sauter mean diameter at steady state: infinite dilution, concentration ϕ, m
F(V,t), F(Z) ,	cumulative distribution function
f(V,t)	over all cumulative distribution function
g	gravitational acceleration constant, m/s^2
g(V/V')	function defined in Equation 2
G(V,V'), G(Z)	cumulative distribution function for drop fragments
H	depth of the upper liquid phase, m
j	the number of bubbles that have entered the upper liquid phase
K, K'	a dimensional function, dimensional coefficient, s^{-1}
L, L'	average drop-drop distance, drop-drop distance, m
n	dimensionless coefficient
P_2, P_3, P_4	coalescence probabilities
T	boundary layer thickness, m
t, t_i^*	time, apparent age of the drop distribution i, s
Ug, U_b	superficial gas velocity, bubble velocity, m s^{-1}
V, $V_{vs,t}$, V_{vs}	drop volume, mean volume based on sauter mean at time t, and at steady state, m^3
v	volume of liquid entrained per bubble, m^3
X_i (t)	the fraction of the liquid entering with bubble i remaining in the upper phase at time t
Y	the initial volume of the upper liquid phase
Z	similarity variable
Γ(V)	transition probability function
μ_c	viscosity of continuous phase, Kg $m^{-1}s^{-1}$
μ_d	viscosity of dispersed phase, Kg $m^{-1}s^{-1}$
ε, ε_{eff}	energy dissipation rate, effective energy dissipation rate, m^2 s^{-3}
Φ	dispersed phase volume fraction in the upper liquid phase
τ	shear stress, Kg $m^{-1}s^{-2}$
σ', σ	surface tension: gas continuous liquid, liquid-liquid, Kg s^{-2}

μ_c, μ_d viscosity of continuous and dispersed liquid phases, Kg m^{-1} s^{-1}
ρ_c density of the continuous liquid, Kg m^{-3}

References

Coulaloglou, C.A., and Tavlarides, L.L. (1976). Drop size distributions and coalescence frequencies of liquid-liquid dispersions in flow vessels. *Chem. Eng. Sci.*, 32, 289.

Coulaloglou, C.A. and L.L. Tavlarides (1977). Description of interaction processes in agitated liquid-liquid dispersions. *Chem. Eng. Sci.*, 32, 1289.

Das, P., D. Ramkrishna and G. Narsimhan (1987). Effect of mass transfer on droplet break-up in stirred liquid-liquid dispersions. *A.I.Ch.E. J.*, 33, 1899-1902.

Dukhedin-Lalla, L., S. Yushun, J. M. Shaw and P. Rahimi (1989). Phase splitting of complex hydrocarbon mixtures. *Fluid Phase Equilibria*, 53, 415-422.

Fillipov, A.F. (1961). On the distribution of the sizes of particles which undergo splitting. *Theory Prob. Applic.*, 6, 275.

Hatate, Y., S. Mori, S. Okuma and Y. Kato (1976). Drop size in gas-liquid-liquid system bubble columns. *Kagaku Kogaku RonBunshu*, 2, 133.

Hatzikiriakos, S.G., R.P. Gaikwad, P. R. Nelson and J.M. Shaw (1990a). Characterization of gas agitated liquid-liquid dispersions. *A.I.Ch.E. J.*, in press.

Hatzikiriakos, S.G., R.P. Gaikwad, P. R. Nelson and J.M. Shaw (1990b). Transitional drop size distributions in gas agitated liquid-liquid dispersions. *Chem. Eng. Sci.*, in press.

Hinze, J.O. (1955). Fundamentals of the Hydrodynamic Mechanism of Splitting in Dispersion Processes. *AIChE J.*, 1, 289.

Kato Y., T. Kago and S. Morooka (1985). Longitudinal concentration distribution of droplets in multi-stage bubble columns for gas-liquid-liquid systems. *J. Chem. Eng. Japan*, 18, 154.

Kondur, R. and J. M. Shaw (1990). The Behaviour of Large Gas Bubbles at a Liquid-Liquid Interface Part-2:Mechanisms for drop entrainment. (in preparation).

Lee, J. C., P. Tasakorn and A. Belghazi (1984). Fundamentals of drop breakage in the formation of liquid-liquid dispersions. The Formation of Liquid-Liquid Dispersions - Chemical Engineering Aspects. ICE, London, 43.

Madden, A.J. and D. McCoy (1969). Drop size in stirred liquid-liquid systems via encapsulation. *Chem. Eng. Sci.*, 24, 416.

Narsimhan, G., D. Ramkrishna and J.P. Gupta (1980). Analysis of drop size distributions in lean liquid-liquid dispersions. *A.I.Ch.E. J.*, 26, 991.

Narsimhan, G., J.P. Gupta and D. Ramkrishna (1979). A model for transitional breakage probability of droplets in agitated lean liquid-liquid dispersions. *Chem. Eng. Sci.*, 34, 257.

Park, J.Y. and L.M. Blair (1975). The effect of coalescence on drop size distribution in an agitated liquid-liquid dispersion. *Chem. Eng. Sci.*, 30, 1057.

Ramkrishna, D. (1973). Drop-breakage in agitated liquid-liquid dispersions. *Chem. Eng. Sci.*, 29, 987.

Shah, B.H. and D. Ramkrishna (1973). Population balance model for the mass transfer in lean liquid-liquid dispersions. *Chem. Eng. Sci.*, 28, 389.

Shaw, J.M., S.G. Hatzikiriakos and R.P. Gaikwad (1988a). Characterization of gas-liquid-liquid tubular reactors 1. Drop size and drop size distribution. A.I.Ch.E. National Meeting, New Orleans, March 6-12, 9f.

Shaw, J.M., R.P. Gaikwad and D.A. Stowe (1988b). Phase splitting of pyrene-tetralin mixtures. *FUEL*, 67, 1554-1559.

Stamatoudis, M. and L.L. Tavlarides (1985). Effect of continuous-phase viscosity on the drop sizes of liquid-liquid dispersions in agitated vessels. *Ind. Eng. Chem. Process Des. Dev.*, 24, 1175.

Yoshida, F. and T. Yamada (1971). Average size of oil drops in hydrocarbon fermentors. *J. Ferment. Technol.*, 49, 235.

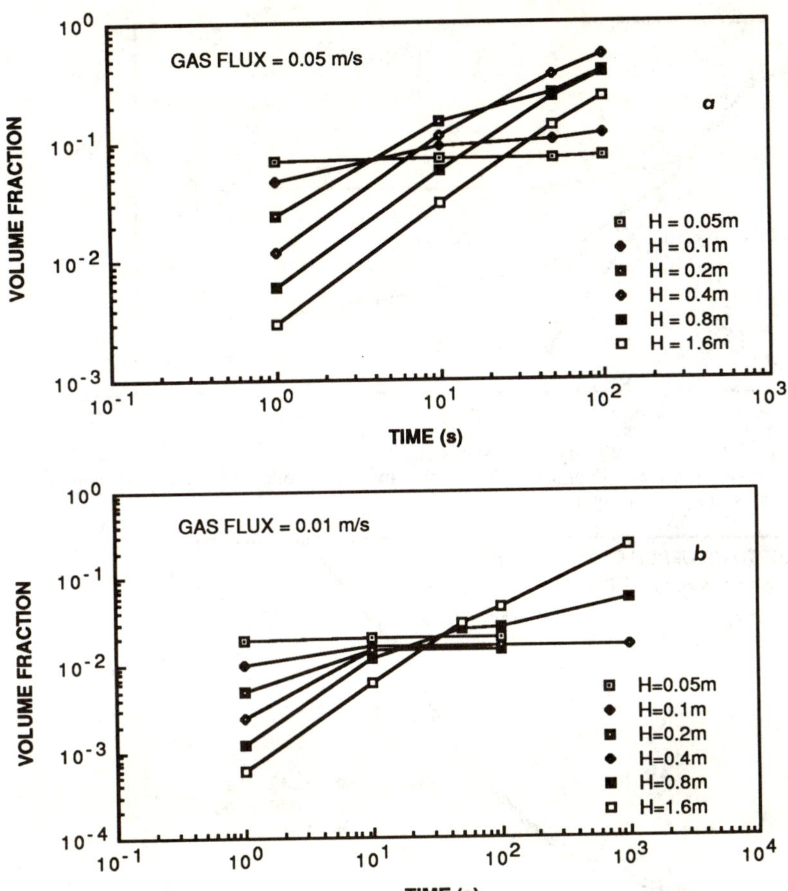

Figure 1: The volume fraction of entrained lower liquid phase material in the upper liquid phase is sensitive to gas flux, the depth of the upper liquid phase, H, and time. Emulsions tend to form at high gas fluxes (a) or long operating times (b).

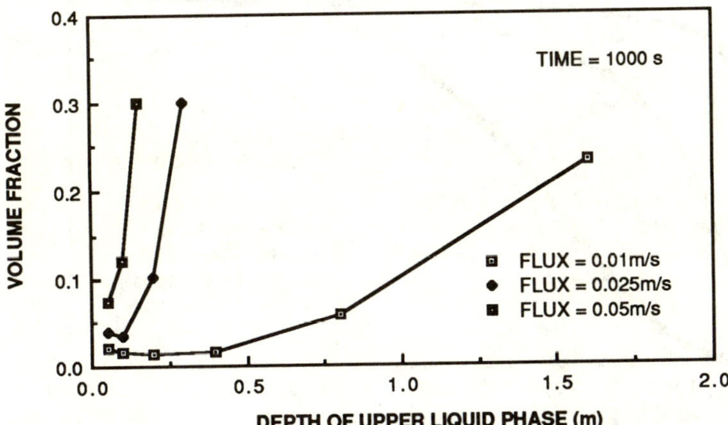

Figure 2: Dispersions tend to become emulsions over time. The rate at which this process occurs is sensitive to gas flux.

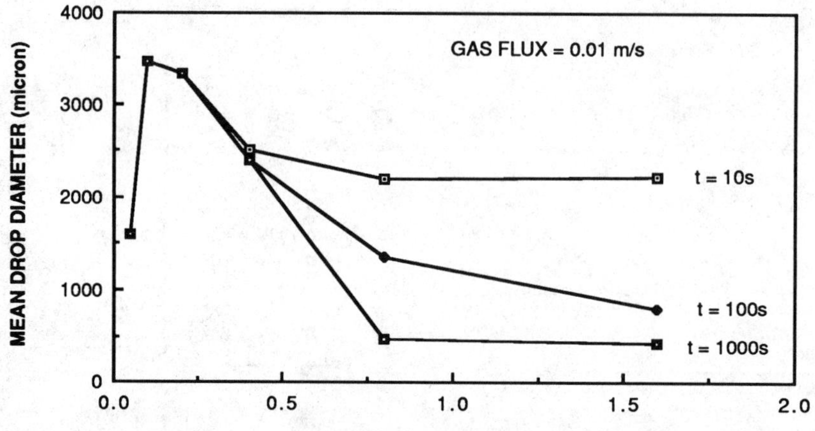

Figure 3: Sauter mean drop diameter exhibits a complex dependence on operating conditions but reduces over time to steady state values.

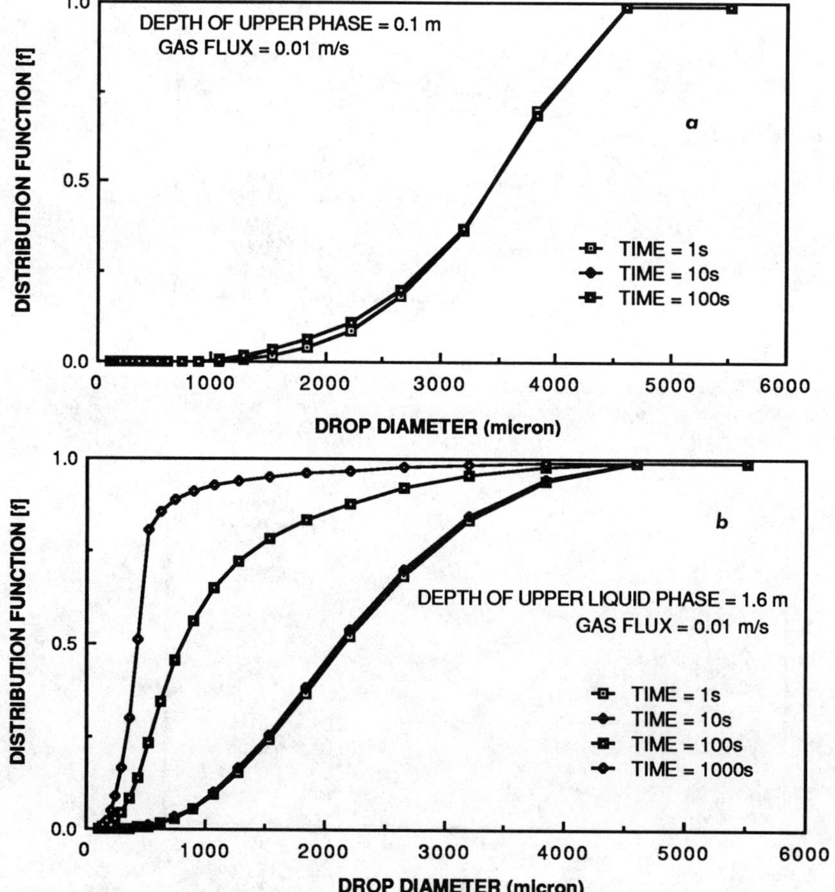

Figure 4: Drop size distributions reach steady state more rapidly in shallow dispersions (a) than in deep dispersions (b).

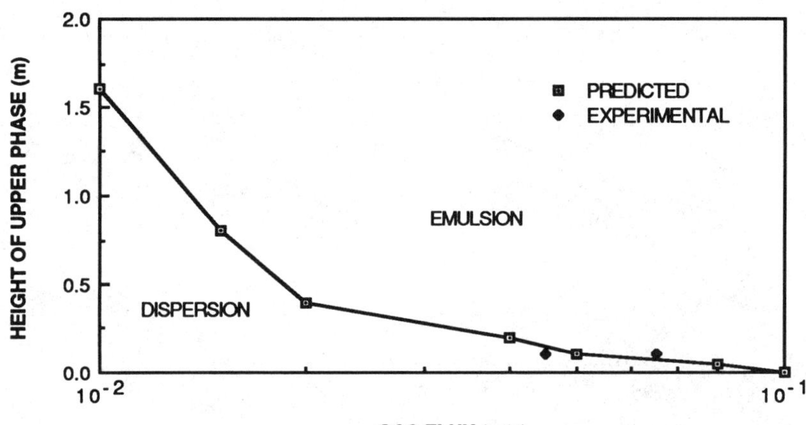

Figure 5: Dispersion stability at 1000s, showing the dependence of stability on the depth of the upper liquid phase. Dispersions can be stable or metastable.

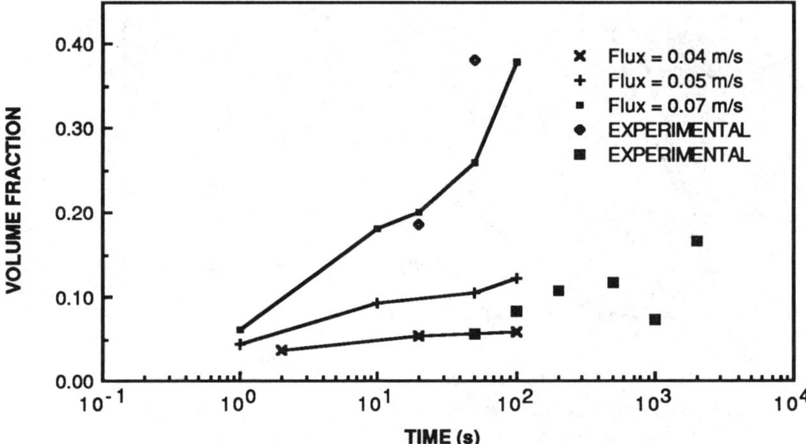

Figure 6: A comparison between predicted and experimerntal concentration profiles.
The fluxes corresponding to the upper and lower experimental data are approximately 0.07 and 0.045 m/s respectively.

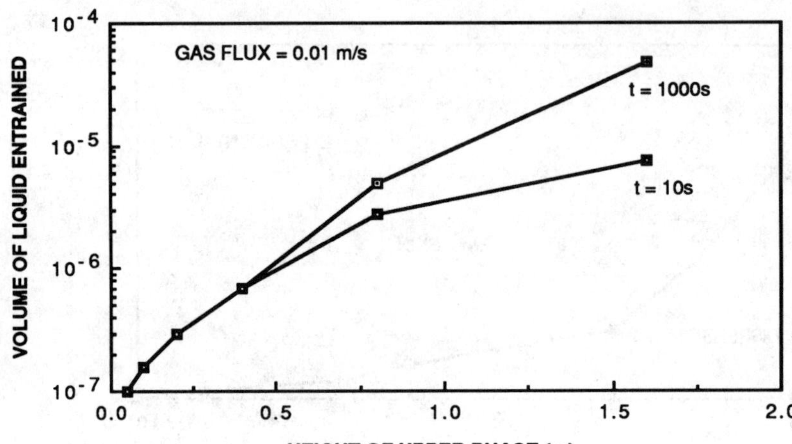

Figure 7: The volume of lower phase liquid entrained in the upper liquid phase is minimized at short operating times and low upper phase depths.

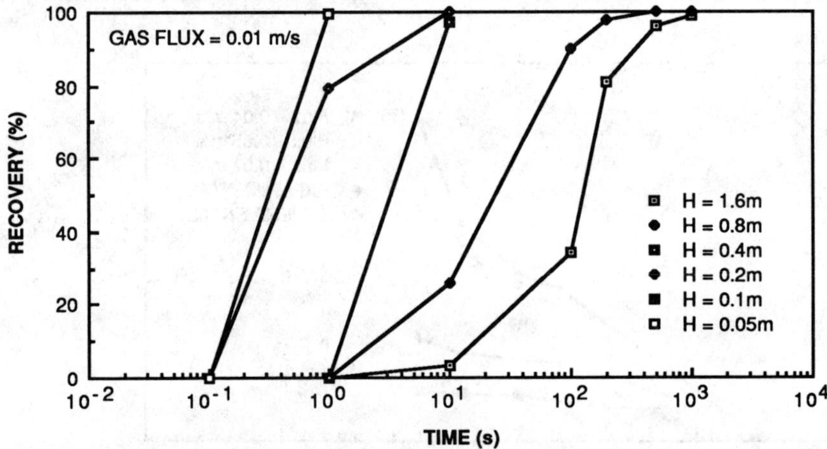

Figure 8: Entrained lower liquid phase material is recovered by the lower liquid phase over time if the gas is stopped. In this example a gas flux of 0.01 m/s was maintained for 1000s.

Unloading and reclaiming bulk solid materials from vessels by pneumatic means

A.A. Reed and A.N. Pittman
The Wolfson Centre for Bulk Solids Handling Technology, School of Engineering Thames, Polytechnic, London, U.K.

Summary

The unloading and reclaiming of bulk solid materials from vessels by vacuum pneumatic pipeline conveying is a technique that has been used by industry for many years. However, whilst these systems work, experience had shown that they are invariably inefficient in operation. Recent research has shown that for free flowing products minor changes in the design of the suction nozzle can have a significant effect on the performance and hence efficiency of such systems. This paper presents and discusses this work. Techniques for reclaiming cohesive, non-free flowing products are also addressed.

1. Introduction

The pneumatic conveying of bulk solid materials through pipelines by negative pressure or vacuum is a technique that has been used by industry since the turn of the century. Today, systems are in operation ranging in throughput from around 1 tonne/hr in small process plant applications to 500 tonne/hr when unloading materials such as alumina, petroleum coke and wheat grains from the holds of ships. However, despite their obvious advantages of flexibility, compact size, few moving parts and cleanliness, such systems tend to be more expensive in terms of energy costs than mechanical techniques. Therefore their use has been limited.

Since such systems operate under negative pressure with respect to the surroundings, they are advantageous in terms of environmental considerations. Therefore, as a consequence of the introduction of stringent regulations in various countries regarding permissible levels of environmental emissions, industry is currently reappraising the vacuum pneumatic conveying system as a viable in-plant transfer facility.

2. System Components

Conventional vacuum conveying systems consist of four essential components: the feeder, the conveying pipeline, the filter separator and the exhauster, Figure 1, and are usually operated with air flow rates that are high enough to cause the product to be transported through the pipeline in a suspension-type mode of flow. Feeders for such systems range from the mechanical types such as rotary valves and screw feeders, which operate on a positive displacement principle, to suction nozzles, which rely in some way on the air induced into the nozzle by the exhauster to entrain the material into the conveying pipeline at the required rate. Clearly, the latter are key components to operations where materials are to be reclaimed from heaps, pits and vessels, the application to which this paper is addressed.

Whilst the operating principles of rotary valves and screw feeders are well understood, the mechanisms by which suction nozzles entrain materials into the conveying pipeline are less clear. Thus, one of the major objectives of this paper is to provide an understanding of the entrainment mechanisms in the commonly employed co-axial tube type of suction nozzle and to demonstrate their effect on the performance of negative pressure/vacuum pneumatic conveying systems.

3. Nozzle Designs and Modes of Operation

Although many different designs of suction nozzle have been employed over the years as feeders for vacuum conveying systems, in terms of reclaiming free-flowing materials from heaps, pits and vessels etc the simple co-axial tube or air sleeve is the type most in widespread use, Figure 2. With this arrangement, the air induced into the system by the exhauster flows down the annular chamber formed by the two tubes and entrains the product into the conveying line at the lower end. As such, there are basically two practical methods of controlling or regulating the flow of material into the conveying line,

i) by varying the position of the inner conveying tube relative to the outer tube at the lower end of the nozzle, and

ii) by throttling the air on inlet to the annular chamber of the nozzle.

3.1 The Effect of Nozzle Geometry

The configuration in which the air sleeve nozzle is commonly used is with the inner conveying tube retracted inside the other tube, Figure 2 i). Under such circumstances it is not unreasonable to assume that the air induced into the annular chamber by the exhauster will take the line of least resistance and therefore entrain very little product into the conveying line. Experience has shown that the only way by which reasonable conveying rates may be achieved with this arrangement is to use, what may be considered by normal pneumatic conveying standards, excessively high air flow rates in relation to the physical size of the system in question. Not surprisingly, this method of feeding material into the pipeline results in a highly inefficient conveying system in terms of energy costs, as well as increasing the likelihood of excessive product degradation and/or wear of pipe bends and associated fittings.

It is clear that, providing the product is sufficiently free flowing, by making small changes to the relative positions of the inner and outer tubes in the entrainment region, it should be possible to obtain a more effective method of feeding material into the conveying line, Figure 2 ii). In this respect work of British Coal in the UK (Refs 1 and 2) has demonstrated that the relative position of the lower ends of these tubes has a significant effect on the conveying rate of 'singles' coal (-25 +13 mm). Their work demonstrates that for a given superficial air velocity at entry to the conveying line, the relationship between the position of the lower ends of the inner and outer tubes and the conveying rate is of the form shown in Figure 3.

Subsequent work by the authors with alumina and other materials has confirmed the general form of this relationship (Refs 3, 4 and 5).

The experimental results from this work on a small scale test rig with a sandy grade of alumina are shown graphically in Figure 4. This shows the effect of the position of the inner tube relative to the outer tube, denoted by the symbol X, for the suction nozzle in an unthrottled condition and over entrainment velocities in the range 20-60m/s. The entrainment velocity represents the superficial velocity of the air on entry to the conveying line.

From this figure the significant effect of nozzle geometry in the range X = -25mm to X = 25mm on the conveying rate is clear, and is of a similar nature to the relationship obtained elsewhere for coal, Figure 3. At values of X greater than 25mm too much product was presented to the pipeline, which resulted in a blocked line. These results also help to explain the earlier comment regarding the necessity for high air flow rates to achieve acceptable throughputs when, as shown in Figure 2 i), the inner tube of an unthrottled nozzle is retracted inside the outer tube.

The reason for the improved system performance with increasing nozzle dimension X in the positive direction was evident from flow visualisation tests. (Refs 3, 4 and 5). This work reveals a marked change in the mechanisms by which material is entrained from the bulk into the conveying line. This confirmed that with the inner tube retracted inside the outer tube, no material was readily presented to the entraining air flow, Figure 5 i). With this particular configuration the conveying air is required to entrain material from a static bed beneath the nozzle as it flows from the annular chamber into the conveying line. Under such circumstances high air flow rates are necessary to achieve reasonable throughputs. Such air flows serve to aerate the product sufficiently in the region beneath the nozzle to enable it to be removed from the bed of material and entrained into the conveying line.

By gradually increasing the value of X in the positive direction the amount of material presented to the entraining airflow will increase, and so the throughput also increases. Ultimately a stage will be reached where the angle between the inner and outer tubes will be greater than the angle at which the material flows onto the annular chamber, Figure 5 ii). Under such circumstances the system is effectively self feeding in that, providing the product is sufficiently free flowing, it will always flow under the influence of gravity into the induced airstream and will then become entrained into the conveying line. Since gravity is, in effect, assisting the feeding process in this case, it is not surprising that this entrainment mechanism is more efficient than with the inner tube in a retracted condition and this is confirmed by the experimental results of Figure 4.

If this reasoning is accepted, it is evident that is not the linear dimension X that is important with respect to the efficient operation of the nozzle and subsequent conveying system, but rather the angle between the inner and outer tubes in relation to the angle at which the product flows into the area beneath the annular chamber. Obviously, this only applies when the nozzle is being used in the vertical direction. Should the nozzle be inserted into the material at some other angle, a different situation may exist, in that it is possible that the induced air will take the line of least resistance through the nozzle and hence the throughput may decrease. The emphasises the point that to ensure maximum throughputs, care should be taken to ensure that this type of nozzle is maintained in the bed of material in the vertical direction.

3.2 The Effect of Throttling

The second method of regulating the flow of material into the conveying line relies on varying the area of flow for the induced air at the inlet to the annular chamber of the suction nozzle, Figure 2 iii). In fact, this approach is probably the most common way of regulating the product conveying rate of systems using this particular type of feeder (Refs 6 and 7). However, it is clear that there is a considerable amount of misunderstanding regarding how this approach actually controls the product flow rate. It is commonly thought that this is the result of regulating the air flow rate and, therefore, varying the ratio of solids to air in the pipeline. This is also the explanation given in a British Standard relating to terms used in pneumatic handling (Ref 8).

In fact, the extent to which varying the flow area on inlet to the nozzle alters the air flow rate induced into the conveying system will depend on the characteristics of the type of exhauster being used to create the operating air flow. For example, should the prime mover be a fan operating on the principle of aerodynamic compression, the general shape of the performance characteristics of these machines means that the added resistance to flow, resulting from decreasing the flow area on inlet to the nozzle, may decrease the air flow rate induced into the system by a considerable amount, Figure 6. As seen in the previous section, the action of decreasing the air flow rate to the system generally decreases the throughput.

Conversely should the prime mover be a positive displacement exhauster of the Roots type, the relatively steep relationship between air flow rate and operating vacuum for such machines means that the added resistance resulting from decreasing the flow area will, by comparison with aerodynamic exhausters, have a small effect on the air flow rate. From this it is evident that whatever type of prime mover is being used, decreasing the area of flow for the air induced into the suction nozzle with a view to regulating the air flow rate into the system will not, by itself, have a beneficial effect in increasing the amount of material entrained into the conveying line. Therefore, if this is accepted, it is clear that some other dominant mechanism must be present to explain the effectiveness of this technique.

In this respect work by the authors (Refs 5 and 9) has shown that the effectiveness of this technique is, in fact, attributable to the decrease in flow area providing a throttling effect on the air as it is induced into the annular chamber. This creates a decrease in pressure of the air flowing down the annular chamber compared with the pressure of the air in the interstices of the product surrounding the nozzle and, as such, leads a physical movement of materials towards the entraining airstream. This results in more material being entrained by the air flow, thereby increasing the product flow rate.

The experimental results demonstrating this effect are shown graphically in Figure 7. The three graphs represent different basic nozzle configurations ranging from the condition with the inner tube retracted inside the outer tube by 20mm (X = -20mm), the tubes level (X = 0) and the inner tube protruding by 20mm (X = 20mm). Each graph demonstrates the effect of throttling on the resultant conveying rate of alumina, for entrainment velocities in the range of 20-60 m/s. The percentage throttle indicated on the bottom axis of each graph is simply a measure of the actual area of flow available to the air on inlet to the annular chamber of the nozzle, in relation to the maximum area available corresponding to the zero throttle condition.

For all three graphs the significant effect of throttling the air on inlet to the annular chamber over the wide range 0 - 90% on the resultant conveying rate is clear. However, the effect is more marked with the basic nozzle configurations corresponding to X = -20mm and X = 0. This is because, as shown in Figure 4, the condition corresponding to X = 20mm at zero throttle is already an efficient feeding configuration. The reason why, for the basic nozzle configuration of X = -20mm, throttling effectively lifts the product into the ntraining airstream may be better appreciated by referring to Figure 8. This demonstrates the effect of throttling on the measured static pressure of the air as it flows down the annular chamber of the nozzle arranged with a basic configuration of X = -20mm. In essence, for a given entrainment velocity, increasing the degree of throttle increases the partial vacuum of the entraining airstream as it flows down the annular chamber. Since the pressure of the air in the interstices of the product surrounding the nozzle will be close to atmospheric pressure, the difference in pressure in this region will then result in a force that effectively lifts or elevates the material in the direction of the pressure gradient. By this way material is then presented to the entraining airstream. This explanation was confirmed by flow visualisation tests. (Refs 5 and 9). Whilst this is a plausible explanation for suction nozzles that have the inner tube retracted inside the outer tube Figure 7 i), for nozzles of the reverse geometry, Figure 7 iii), the increase in conveying rate with increasing throttle is probably due to a combination of material being elevated from the static bed beneath the nozzle, together with an increased quantity of material being presented to the area beneath the annular chamber.

4. Discussion

In the foregoing sections the principal mechanisms of entrainment of free-flowing materials in co-axial tube or air sleeve suction nozzles and their effect on the performance of vacuum pneumatic conveying systems have been identified and discussed in detail. These may be summarised as:

i) With the nozzles in an unthrottled condition and with the inner conveying tube retracted inside the outer tube, the air induced into the nozzle by the exhauster takes the line of least resistance and, therefore, entrains little product into the conveying line. Under such circumstances high air flow rates are necessary to achieve reasonable conveying rates, Such air flows serve to aerate the product sufficiently in the region beneath the nozzle to enable it to be removed from the bulk and entrained into the conveying line. Not surprisingly, this entrainment mechanism is very inefficient in terms of overall system performance.

ii) Changing the nozzle geometry such that the angle between the protruding inner tube and outer tube is greater than the angle at which the product flows into this region leads to a more efficient entrainment mechanism than that outlined in i). Under such circumstances the system is self feeding, in that, providing the product is sufficiently free flowing, it will flow under the influence of gravity into the airstream induced into the annular chamber by the exhauster and then becomes entrained into the conveying line.

iii) Throttling the air flow on inlet to the annular chamber of the suction nozzle is an effective method of improving the mechanism of entrainment, irrespective of the basic nozzle geometry. This is attributable to the throttle decreasing the pressure of the air flowing down the chamber, in relation to the pressure of the air in the interstices of the product surrounding the nozzle, thereby causing a movement of material towards the entraining airstream. The results in more material being entrained by the induced air flow, which, in turn, is reflected in an increase in product flow rate through the system.

From the above it is evident that should insufficient thought to be given to the design and operation of suction nozzles as the feeder for vacuum pneumatic conveying systems, the conveying capacity of the pipeline, in relation to the quantity of air induced into the system by the exhauster, may well be under-utilised. By understanding the potential entrainment mechanisms in such nozzles, it can be seen that it may be possible to enhance their performance as a feeder and thereby improve the effectiveness of the conveying system as a whole. However, a word of caution is necessary regarding using this approach to improve the performance of existing systems that are currently using an inefficient entrainment mechanism. This caution is related to matching the capabilities of the nozzle to feed material into the pipeline at the required rate, the conveying potential of the pipeline itself and the exhauster that is being used to generate the operating air flow. Should the exhauster be a Roots type machine, then varying the basic nozzle configuration or degree of throttle will have little effect on the air flow rate. Therefore, providing the exhauster is not operating at the limit of its pressure capabilities, either of the approaches identified in ii) and iii) may be used to enhance the overall performance of the system. Conversely, should the prime mover be a fan operating on an aerodynamic principle, then increasing the amount of material introduced into the pipeline by employing either of these approaches may increase the operating pressure of the fan by a considerable amount. As shown in Figure 4, the general shape of the operating characteristics of this particular type of machine means that the air flow rate induced into the system may well reduce to a value below that necessary for successful conveying.

4.1 Cohesive Products

From the foregoing discussion it has been seen that the effect of suction nozzle operation on the conveying rate of free flowing materials through vacuum conveying systems is significant. Clearly, there are many materials which, for a variety of reasons are not free-flowing, and therefore will not self-feed under the action of gravity to the suction nozzle of such systems. For the efficient conveying of such materials it is clear that the product should be rendered more free flowing in the region of the suction nozzle. Two techniques to achieve this are employed. The first relies on the application of air to the material contained in the feed vessel and the second is based on mechanically stirring the product in the region of the suction nozzles. By using such approaches the implications of the previous sections will still apply to the conveying of such materials.

5. Concluding Remarks

From the foregoing discussion it is clear that poor design and operation of suction nozzles can prevent the full potential of vacuum conveying systems for reclaiming materials from vessels etc from being realised. In this respect it is hoped that this paper has provided the reader with a better appreciation of how these important, but often neglected factors can improve the entrainment mechanisms of materials in suction nozzles, thereby improving the effectiveness of such conveying systems.

References

1. Kaye, W G. Coal and ash handling in industrial boilerhouses, Proc, Conf. on Coal Handling and Coal Qualities, Dansk Ingeniorforeing, Copenhagen, Denmark (1981).

2. Foster, C. The development of a pneumatic suction coal conveying system for boilerhouses, Proc. 1st. Int. Conf. Pneumatic Conveying Technology,

Stratford-upon-Avon, UK (1982).

3. Reed, A R. and Mason, J S. The effect of suction nozzle design on the performance of vacuum pneumatic conveying systems', J. Powder and Bulk Solids Technology, 7, 4, 9-11 (1983).

4. Reed, A R. and Mason, J S. The effect of suction nozzle geometry on the performance of pneumatic ship unloaders, J Pipelines, 5, 67-75 (1985).

5. Reed, A R, Pittman, A N. and Mason, J S. Entrainment mechanisms in co-axial tube suction nozzles and their effect on the performance of vacuum pneumatic conveying systems, Proc. 2nd. Int. Conf. on Bulk Materials Storage, Handling and Transportation, Wollongong, Australia (1986).

6. Grimshaw, J. Vacuum handling, Proc. IMechE Conf. Vacuum Technology, London, UK (1965).

7. Massey, H C. Pneumatic handling techniques for bulk commodities, Proc. 1st Bulk Handling and Transport Conf. Rotterdam, The Netherlands (1977).

8. Anon, British Standard. 3810: Glossary of terms used in materials handling. Part 3. Terms used in connection with pneumatic and hydraulic handling (1967).

9. Reed, A R. and Pittman, A N. The effect of suction nozzle operation on the performance of vacuum pneumatic conveying systems. Proc. 10th. Int. Powder and Bulk Solids Handling and Processing Conf. Chicago, USA (1985).

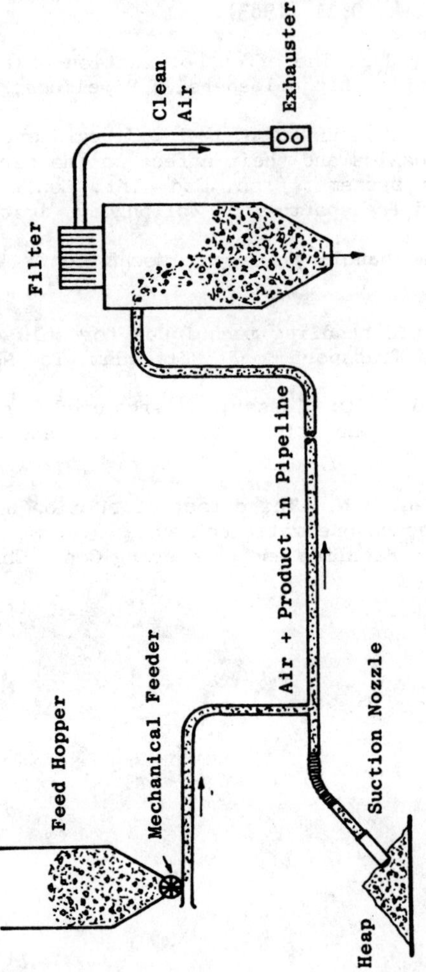

Figure 1 Salient features of negative pressure pneumatic conveying systems

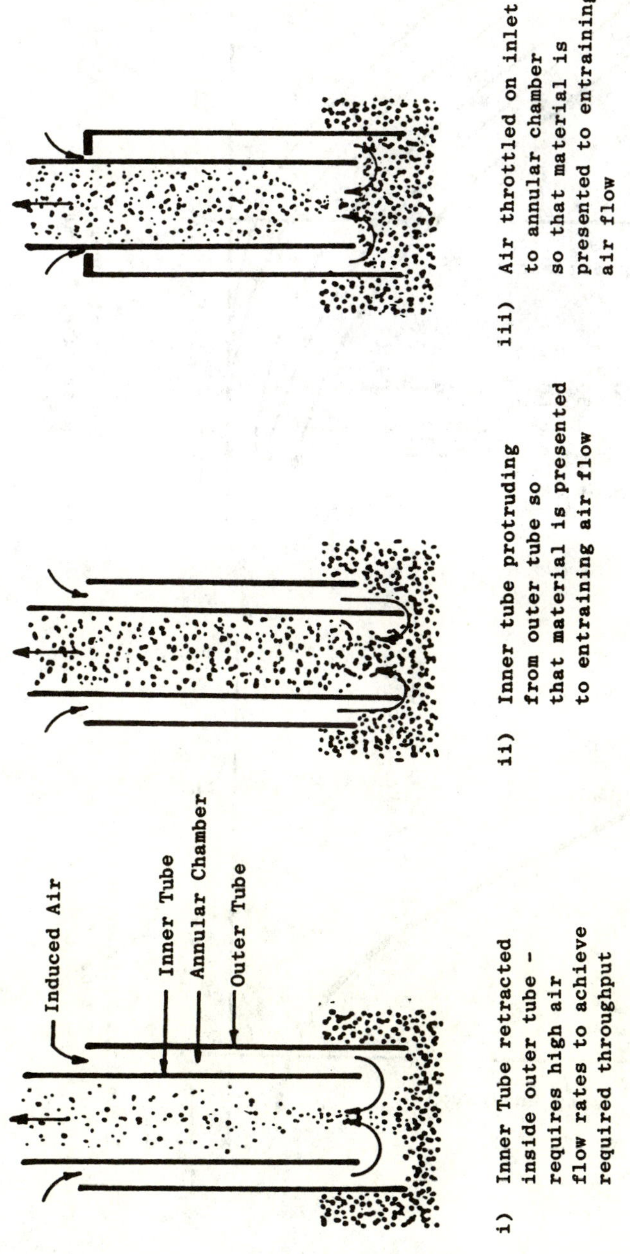

i) Inner Tube retracted inside outer tube - requires high air flow rates to achieve required throughput

ii) Inner tube protruding from outer tube so that material is presented to entraining air flow

iii) Air throttled on inlet to annular chamber so that material is presented to entraining air flow

Figure 2 Different modes of operation of the co-axial tube type of suction nozzle

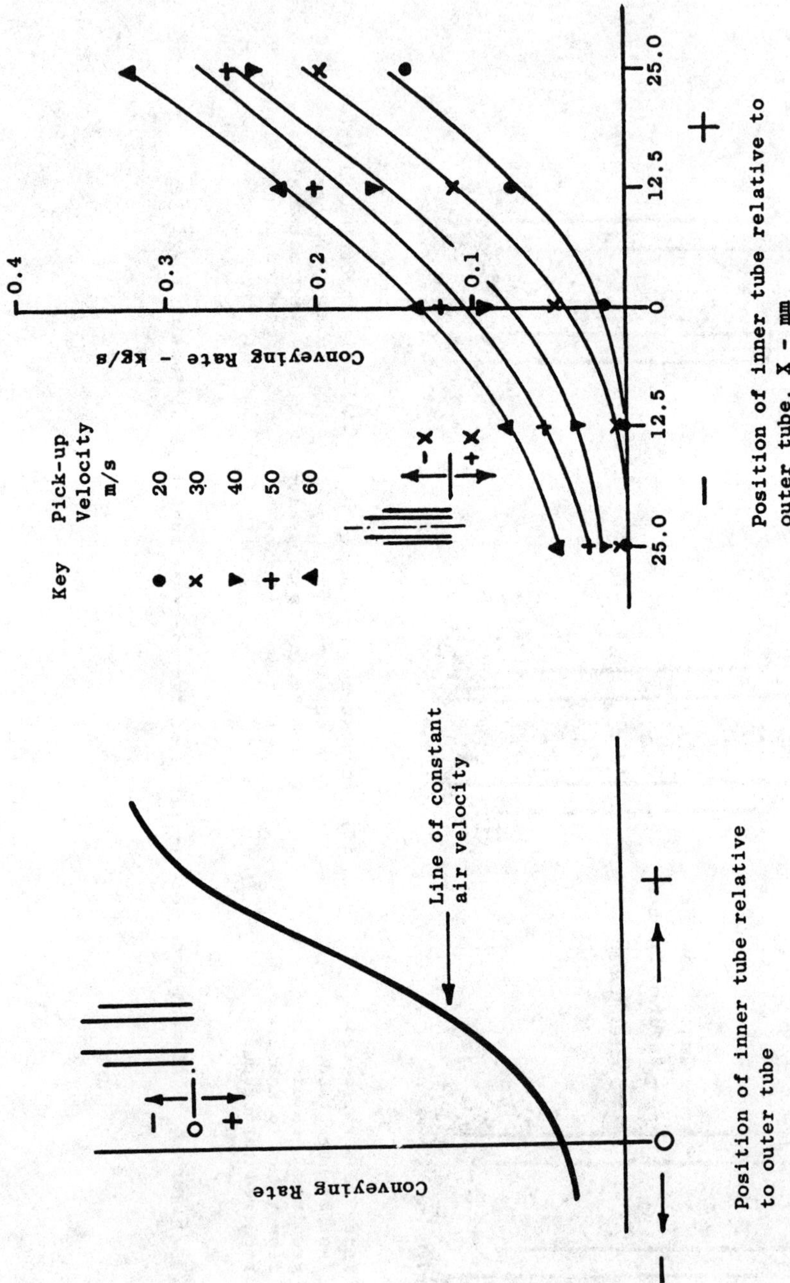

Figure 3 Effect of geometry of nozzle in entrainment region on conveying rate of coal

Figure 4 Effect of nozzle geometry on conveying rate

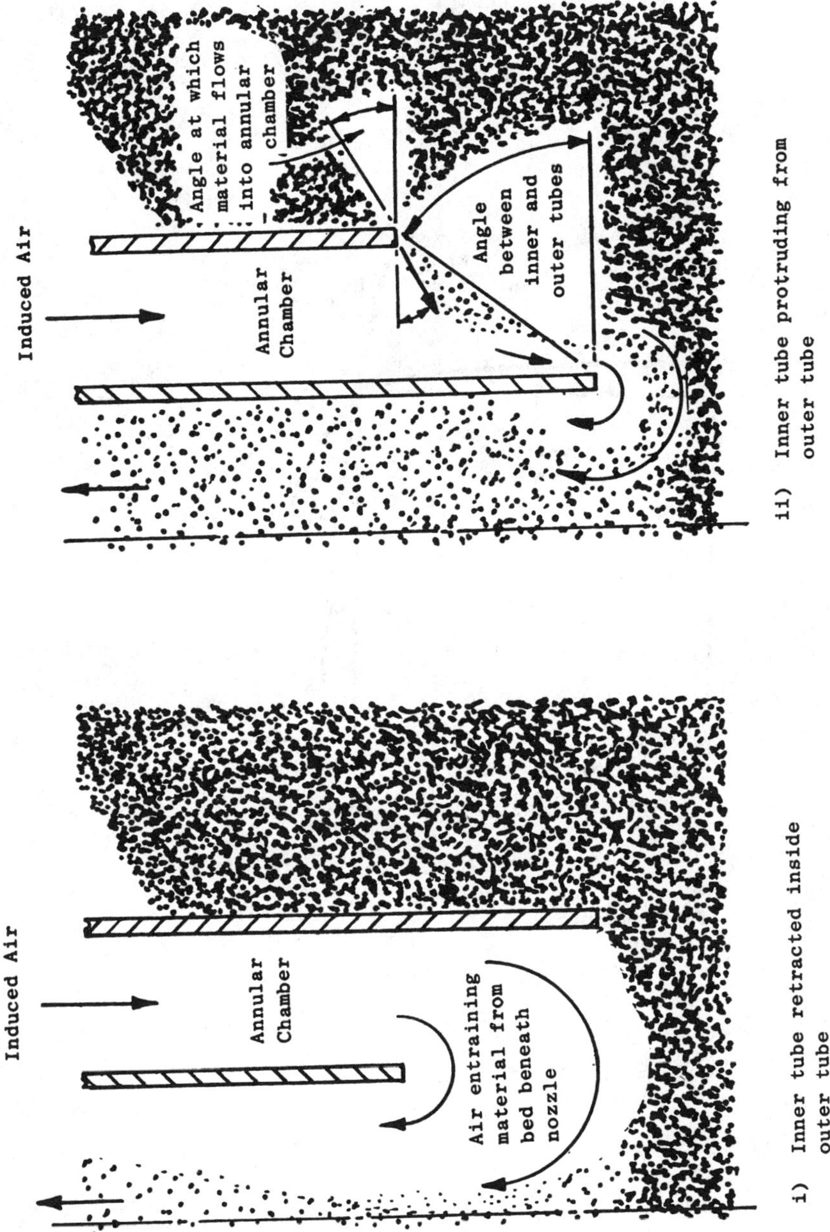

Figure 5 Entrainment mechanisms for different nozzle geometries

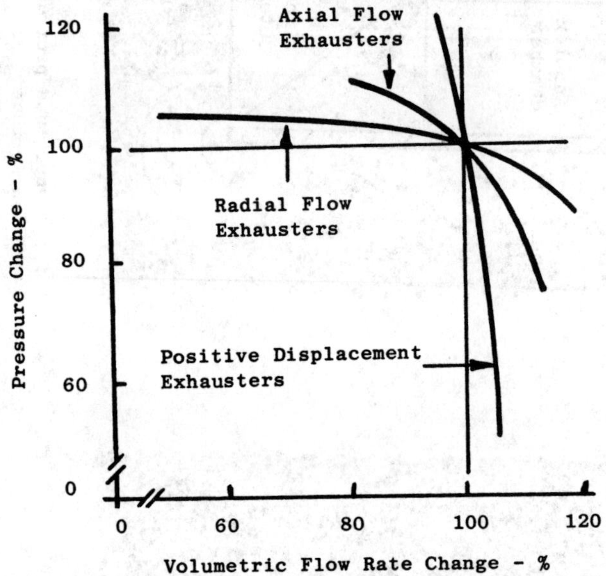

Figure 6 Constant speed characteristics of aerodynamic and positive displacement exhausters

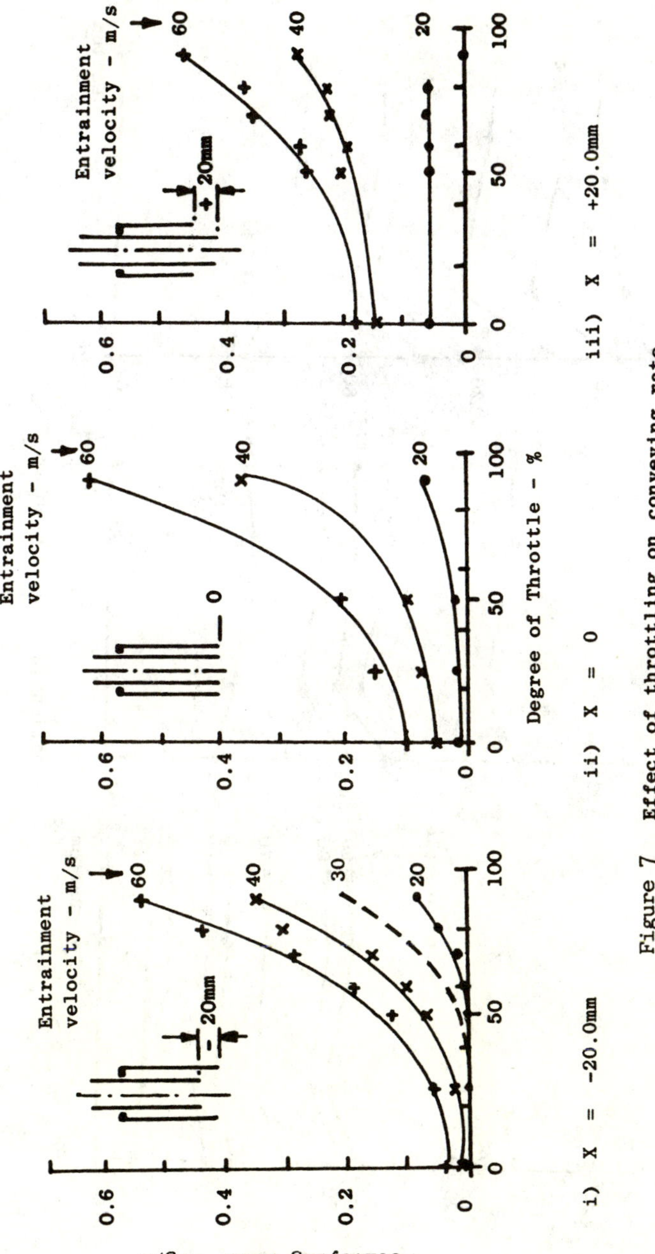

Figure 7 Effect of throttling on conveying rate

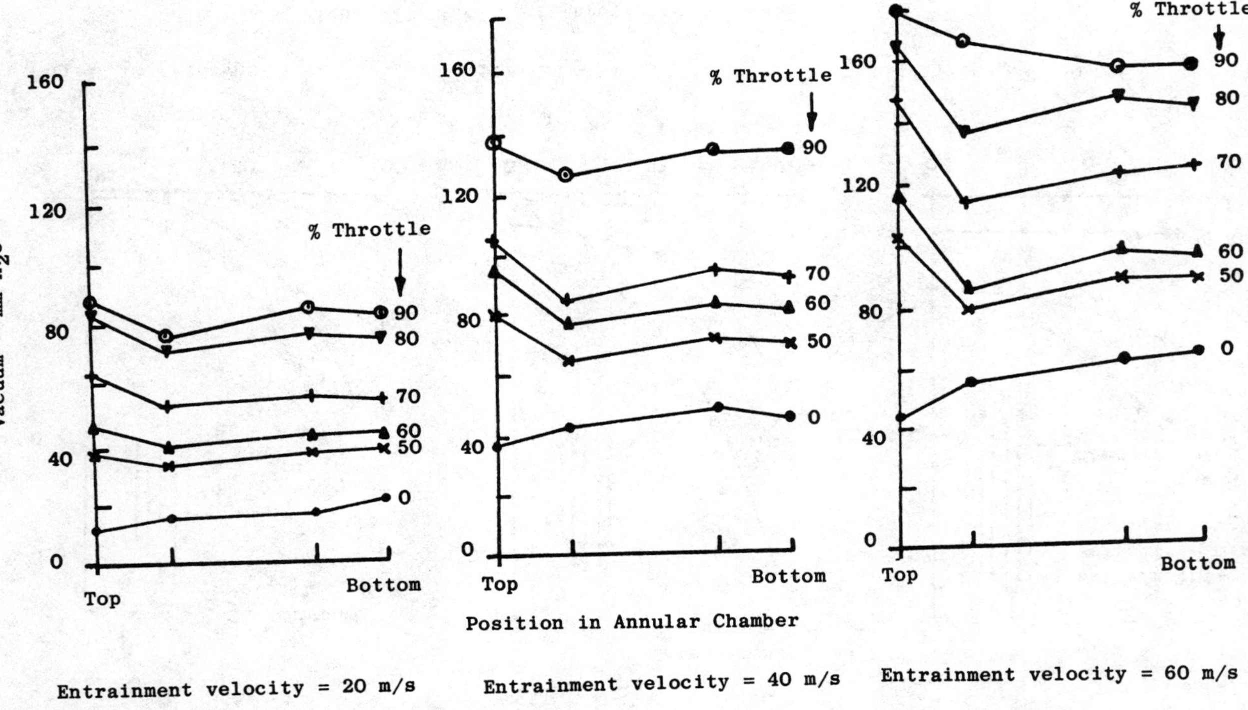

Figure 8 Effect of throttling on the pressure distribution in the annular chamber - for a nozzle geometry X = -20mm

Pneumatic conveying for the metals industries, conveying by kilogram

G.C. Cooper
Simon-Macawber Limited, Doncaster, England

Summary

This paper discusses a pneumatic conveying system which has been developed for specific needs of various process industries. The Simon-Macawber injector provides accurate material dosing on a continuous or batch basis. Can overcome process back pressure, and handle a wide range of material sizes and densities.

The operating principles, as well as case studies showing typical applications are included in this paper.

Introduction

It has long been recognized that there is a requirement to feed material continuously, smoothly and accurately in various industrial processes. To satisfy this requirement, an injector was developed which enables accurate dosing of materials at relatively small flow rates and over moderate distances into industrial processes.

This system has the advantage over a low pressure lean phase system of being a self-contained pressure vessel able to use high pressure gas for conveying to overcome process backpressures. Also, due to the mechanical method of feeding materials out of the vessel a much wider range of injectants can be handled as opposed to the uncertainty of pressure differential induced flow.

Conveying By Kilogram

The Simon-Macawber injector comprises of a lock hopper and injector vessel complete with metering feeder, associated valve controls, and manifolds to form an independent unit (Figure 1).

The lock hopper cycles to take in material from a supply hopper located above, pressurizes it to conveying pressure and transfers the material into the injector vessel so that continuous injection is maintained. The lock hopper cycle is automatic, triggered by a level probe in the injector vessel.

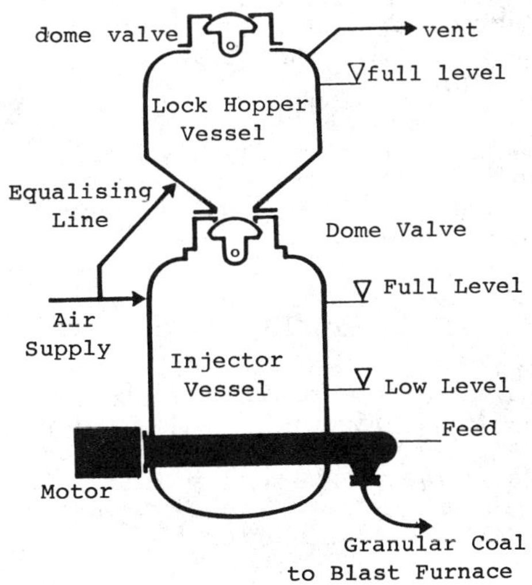

Conveying gas, normally compressed air, enters the injector vessel continuously to maintain a constant pressure within. The metering feeder, located in the base of the vessel dispenses the injectant into the conveying air-stream with a volumetric accuracy within +/- 2% of maximum output. The feeder is driven by a standard electric motor and gearbox. Motor speed and therefore, feeder output is controlled by a variable frequency drive unit which permits a 10:1 turndown ratio.

The output of one injector may be divided to more than one reception. Two methods are available. A pipeline flow splitter is available which divides the flow without the use of moving parts to permit an injector to feed two points. Alternatively, a motorized splitter is available with up to 18 outlets, the number depending on the total number of feed points to be served.

Injectors have been applied to many processes and for a wide range of injection rates. The smallest is a mobile unit used for treatment of molten aluminum in die-case plant holding ladles. Its delivery rate range is 20-200 gr/min. The largest unit, located in India, injecting pulverized coal into power boiler start-up burners for flame stabilization. Each has an output of 12 tons of pulverized coal/hour (200 kg/min).

To further illustrate the metallurgical applications of the injection system following are three brief descriptions of injector applications.

Case Study - British Steel Plc

It would not be possible to talk about injection and Simon-Macawber without mentioning granular coal injection into blast furnace tuyeres. British Steel Plc is our largest user of injectors with, to date, 132 units installed on three sites.

Coal is injected into the blast furnace for a number of reasons, not least of which is that coke at the bell is approximately twice the cost of coal at the tuyere. For the coke short ironmaker who is forced to purchase coke on the open market this represents a considerable saving on iron production cost, typically $3-5/tonne.

The preparation and injection system for coal represents only 10-15% of the investment cost of a coke oven with the equivalent fuel output.

Injecting coal at the tuyere does not require special coal properties. The following table shows a wide range of coal volatility, ash and sulphur contents that have been successfully injected. Metallurgical coal is not required.

TABLE 1
Range Of Coal Injected - Chemical Analysis

	% As-Injected			% Dry Basis						
Coal	Total H_2O	Inherent H_2O	Surface H_2O	V.M.	Ash	S	C	H_2	N_2	O_2
1	7.7	-	-	38.6	3.4	1.6	80.9	5.2	1.6	7.1
2	3.5	3.1	0.4	26.9	6.7	1.0	81.5	4.6	1.5	4.5
3	3.0	2.4	0.4	26.8	11.9	0.8	76.7	4.4	1.4	4 6
4	5.1	4.2	0.9	30.4	9.0	1.3	78.7	4.4	1.5	4.8
5	3.9	2.6	1.3	36.0	5.2	1.4	78.8	5.1	1.5	7.7
6	2.0	1.6	0.4	31.1	3.4	1.5	78.6	4.8	1.5	7.1
7	1.4	0.4	1.0	22.1	10.8	0.3	79.0	4.3	1.4	4.1

For most operations the raceway flame temperature (RAFT) is an important measure and indicator of the furnace fuel balance. Injection of auxiliary fuels into the tuyere is an endothermic reaction. However, (see Table 2) coal has the lower temperature depressing effect. Therefore, more coal can be injected than for oil or gas injection, thus, more coke can be replaced.

Optimization of flame temperature provides the potential for minimizing fuel rate, and maximizing productivity. The lowest coke rates on any furnace have been achieved with coal injection.

TABLE 2
Effect of Injectants Upon Blast Furnace

Parameter	Change	Change in Flame Temperature Deg. C	Change in Coke Rate kg/thm
Tuyere Injection			
Anthracite Coal	+ 100 kg/thm	-162	-91.0
High Volatile Coal	+ 100 kg/thm	-218	-76.1
Heavy Fuel Oil	+ 100 kg/thm	-321	-98.4
Natural Gas	+ 100 kg/thm (+ 132 Nm3/thms)	-513	-82.6
Blast			
Humidity	+ 10 g/Nm2	-58	+ 7.7
Temperature	+ 100 Deg.C.	+ 83	-15.5
Oxygen Enrichment	+ 1%	+ 54	-1.4

When considering high coal injection rates - above 150 kg/thm, the ability to control coal flow to each individual tuyere becomes very important. Variations in distribution of coal and hot blast around the furnace result in variation of flame temperature, coke consumption and tuyere gas generation. In order to minimize these variations it is necessary to comparatively measure the blast flow to each tuyere and match the coal injection rate to the blast flow on a tuyere by tuyere basis. A Simon-Macawber injector, applied to each blast furnace tuyere, provides this level of control.

The application of an injector for each tuyere also provides a high level of operating redundancy. If 2 or 3 injectors or tuyeres are unavailable for injection for any reason, the required fuel rate can be maintained by increasing the output of the remaining machines. British Steel's reported availability to the injection system is 99.8%.

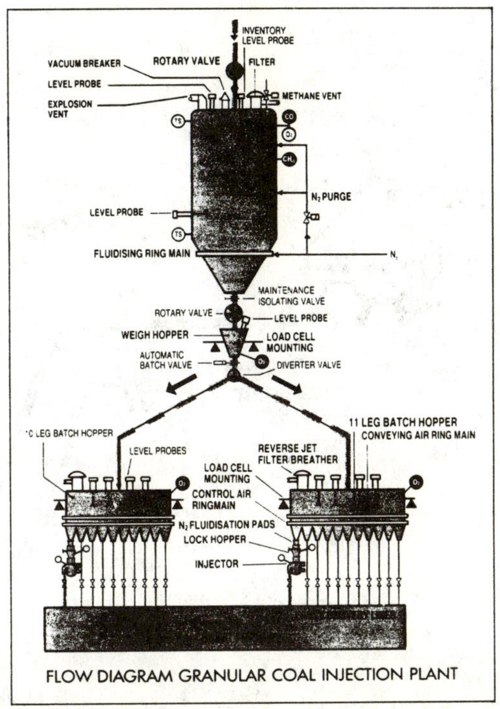

FLOW DIAGRAM GRANULAR COAL INJECTION PLANT

High coal injection rates usually require relatively high percentages of oxygen enrichment, provided through additional lances at each tuyere. The oxygen available from the blast air can be individually supplemented by oxygen enrichment to match the coal injected in each tuyere to obtain optimum raceway conditions. A full scale experimental system to test oxy coal injection to its limits has been built at British Steel Cleveland Works with funding from the European Coal and Steel Community, Hoogovens and Italsider. The system has been designed to achieve an injected fuel rate of 400 kg/thm, double the current maximum in practice.

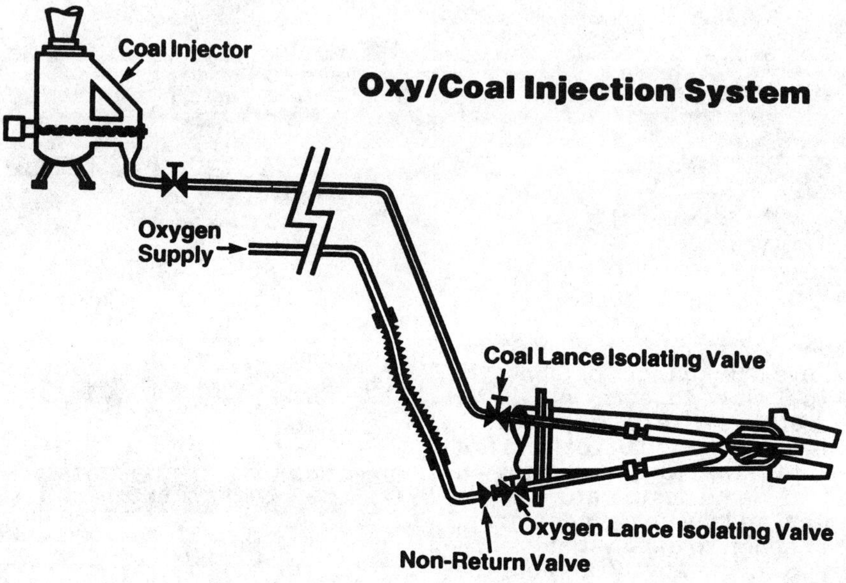

Blast furnace coal injection is not an easy application. Existing iron works were not laid out with coal injection in mind. The available space to place an injection system is often some distance from the furnace, thus conveying distances can be quite long, up to 210 meters in the case of Ravenscraig. The tuyere environment where the coal is discharged is hostile, with hot blast gases at 1000-1200°C traveling at about 180 m/sec into the furnace at a pressure of 1.5-3.0 Barg. Even with these difficulties over 1.5 million tonnes of granular coal have been injected into British Steel furnaces during the last six years, with a reliability (availability) of better than 99%. British Steel Plc, our partners during system development, are also now partners in marketing the technology throughout the world.

Case Study - Arc Furnace Dust Treatment

The second case is based around the controlled injection of electric arc furnace (EAF) dust into a plasma arc treatment facility.

Environmental legislation has classified EAF dust as a hazardous waste. Finding the complete solution for dust treatment is a significant problem for the Iron and Steel Industry. Davy McKee, Ltd. (of Stockton) has developed and demonstrated a process which converts EAF dust into a non-hazardous slag and allows recovery of metals in a saleable form.

The process utilizes the Davy "Hi-Plas" furnace which features a proprietary sleeve reactor surrounding a long D.C. transferred plasma arc. Dry EAF dust, coke and flux are mixed and placed into a Simon-Macawber continuous injector for dosing tangentially into the sleeve at four points above the plasma gun tip with sufficient accuracy, uniformity and velocity to form a covering over the inner wall of the reaction sleeve. Radiant and convected heat from the arc melts the reactants to form a falling film of molten material flowing down the inner wall of the sleeve. The molten material then drops into the furnace hearth region where the reaction is completed.

The reactions take place at temperatures of about 1600°C. A non-hazardous slag and the non-volatile metals are recovered in the furnace hearth.

The single controlled discharge from the injector is split evenly into two streams in static, flow divider type splitters. The line to line flow accuracy is $\pm 5\%$. Flow accuracy and turndown ability are of primary importance to match the dust mixture input to the heat input from the arc in order to achieve complete smelting.

Case Study - Cupola Injection System

The final case study refers to an innovative method of injecting various materials into a foundry cupola, or from the same injection system, treating the molten iron in the holding ladle. Though a number of foundries in the U.K. have shown interest in the system, no installation has yet been made.

Building on past experience of coke breeze injection into cupolas and the more arduous coal injection to blast furnaces, and in conjunction with a well known additives supplier, the Cupola Internal/External Injection System was developed.

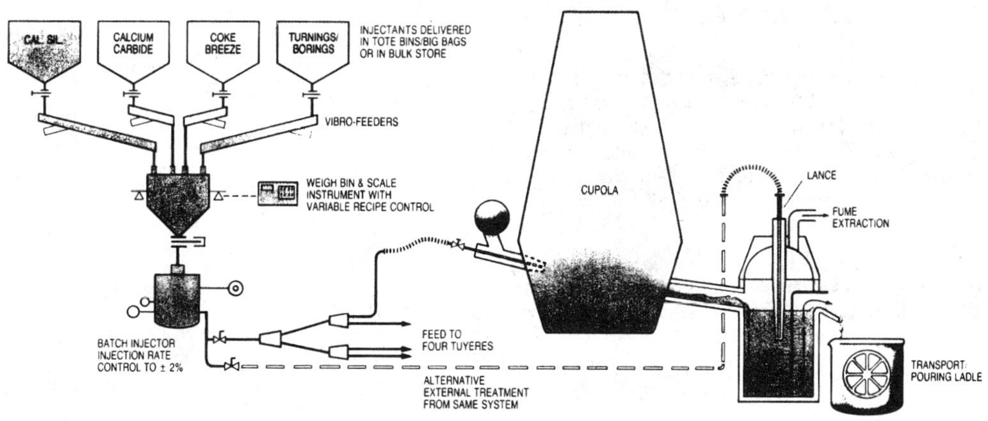

CUPOLA INTERNAL/EXTERNAL INJECTION SYSTEM

Due to the injectors ability to handle a wide range of materials, various additives can be injected through the cupola tuyeres in differing preparations and combinations to achieve the desired iron composition. Any material can be injected, provided it can be processed to a condition suitable for pneumatic injection. The materials in the diagram are turnings and borings which can be injected as an alternative to scrap charging at a considerable cost saving over solid scrap. Typically, up to 20% of the charge can be injected in this way.

Similarly, coke breeze can be injected instead of charging coke, also as a cost saving measure.

Calcium carbide and calcium silicon are examples of additives which can be injected either internally or externally through the same injection system to achieve a wide range of metal chemistries.

Conclusion

The purpose of this paper is to outline Simon-Macawber's capability with regard to continuing pneumatic injection for the Iron and Steel Industries.

The injector is able to dispense and continuously convey a wide range of materials into an equally wide range of processes, with a volumetric accuracy of $\pm 2\%$ against selected rate and with a controlled turndown ratio of 10:1.

Several case studies have been briefly discussed to demonstrate the system and the ability to convey by kilogram for the metals industry.

Pipeline conveyors—an economic alternative for dedicated transport

M. Marcu
Southern Ontario Technologies, Inc., Hamilton, Ontario, Canada L8E 2Y4

Definition-Area of Application

The term "Pipeline Conveyor" was used probably first time by Mr.William Hark in 1976.The Pipeline Conveyors,sometime called Capsular Tube Conveyors or Large Diameter Tubular/Capsule Transport are usually dedicated/shuttle systems using large diameter pipelines or tubular guideways and containers/chasises or platforms to transport different bulk materials on flat,inclined or vertical routes.They could be used also for packaged materials or even people.

The containers involve either wheeled or magnetic suspension (Maglev) and the propulsion could be pneumatic,electromagnetic (LIM) or combined.In the most common variant used today in USSR, Romania,Japan,these systems are like a huge tubular pneumatic post(dispatch),the only difference being that the tube could be 40" in diameter instead of 5" or 6 " and the carriers could carry 6 tonnes instead of a pound.

These systems could be used economically on distances from 1/2 miles to 100 miles and over, for flowrates from 100,000 t/yr to 10,000,000 t/yr and granulations up to 12".

The Elementary Physics of Pneumatic wheeled Propulsion

The propulsion force:

$$F_{P_c} = \pm\, CD \cdot \rho_{air} \frac{(V_a - V_c)^2}{2} \cdot S_c \qquad \text{(Tolmien-Prandtl-1927)}$$

or,

$$F_{P_c} = \pm\, CD \cdot \frac{\Delta Q_{ac}^2}{(\rho_{air\,f} + \rho_{air\,back}) \cdot S_t^2} \cdot S_c \qquad \text{(Marcu-Tomita 1980-1985)}$$

Where:

Fp_c =the propulsion force on a container or on a train of containers;CD=the sealing coefficient of a container or a train of containers;$\pm$ is depending of the magnitude of the air speed and container speed,that is if the air speed is smaller than the container speed then the sign is positive(pneumatic propulsion happens),if the air speed is smaller than the container speed then the sign is negative and in this case the pneumatic braking is happening; ρ_{air} is the density of air; V_a is the air speed; V_c is the container speed; S_c is the cross area of the container used for propulsion(in this case the gasket area); ΔQ_{ac} is the relative massic flowrate(air) passing over the container; ρ_{airf} is the air density in front of the container; $\rho_{airback}$ is the air density in the back of the container; S_t is the cross area of the tube.

Correspondingly the pressure drop associated with the container is:

$$\Delta P_c = \pm CD \cdot \rho_{air} \frac{(V_a - V_c)^2}{2} \quad \text{or} \quad \Delta P_c = \pm CD \frac{\Delta Q_{ac}^2}{(\rho_{airf} + \rho_{airback}) \cdot S_t^2}$$

The relation between airspeed in the tube, container speed and the airspeed in the gap between the container and the tube is given approximately by the following relation:

$$S_{gap} \cdot V_{airgap} \cdot \rho_{airgap} + S_c \cdot V_c \cdot \rho_{airtube} = S_t \cdot V_{airtube} \cdot \rho_{airtube} \qquad \text{(Marcu 1980)}$$

or for pressure drops up to 2 psi(cca 1500 mmWG) and speeds up to 80 m/s:

$$S_{gap} \cdot V_{airgap} + S_c \cdot V_c = S_t \, V_{airtube} \qquad \text{(Marcu-Polizu 1974)}$$

The approximate airspeed in the gap is given by the relation:

$$V_{airgap} = V_{airtube} + k_1 \sqrt{2 \frac{\Delta P_c}{\rho_{airtube}}} \cong V_{airtube} + \sqrt{2 \frac{\Delta P_c}{\rho_{airtube} \cdot CD}}$$

Where: S_{gap} is the area of the gap between container and tube; V_{airgap} is the airspeed in the gap; ρ_{airgap} is the density of air in the gap; $\rho_{airtube}$ is the density of the air in the tube; $V_{airtube}$ is the airspeed in the tube; k_1 is an experimental coefficient dependent on the architecture of the space between the container and the tube.

The equation of motion of the container or a train of containers is:

$$S_c \left(\pm CD \frac{\Delta Q_{ac}^2}{(\rho_{airf} + \rho_{airback}) \cdot S_t^2} \right) - \sum_{i=1}^{2} SO_i (\mu \pm tg\alpha_i + r_{ci}) = (M + mr) \frac{dV_c}{dt}$$

(Marcu 1978)

Where: i is the associate index to a wheel assembly(in case when we have a train); SO_i is the associate load on a wheel assembly; μ is the rolling coefficient; $tg\alpha_i$ is the factor associated with the inclination of the pipeline at the particular point where a wheel assembly (indexed by i) really is; r_{ci} is the curve resistance; M+mr is the total mass of the container(or train) including the reduced mass of the wheels; dV_c/dt is the acceleration of the container.

When the pressure drop is below 1500mmWG(cca 2 psi) and the

air speed up to 20 m/s the air could be treated as an incompressible light fluid with unidimensional rough flow independent of Reynolds number.Also because the shocks are avoided, the transient behaviour of the air could be treated by taking into account the inertial term of the air mass in the tube(which could be for one mile 40" tube about 3000 pounds).In this case, the density of the air could be taken as an average and the air and the containers are governed by the follwing system of differential equations of motion:

$$\begin{cases} \lambda \frac{l}{d} \rho_a \frac{V_a^2}{2} + K\rho_a \frac{V_a^2}{2} + CD_s \rho_a \frac{V_a^2}{2} + CD_i \rho_a \frac{V_a^2}{2} + \sum_{i=1}^{n} \pm CD_c(sc_i) \frac{(V_a - V_{c_1})^2}{2} \rho_a + \cdots + l\rho_a \frac{dV_a}{dt} = \\ \qquad = -AV_a^2 + BV_a + C = f(V_a) \\ \cdots\cdots\cdots\cdots\cdots\cdots\cdots\cdots \\ S_c \left(\pm CD_c(sc_i) \cdot \rho_a \frac{(V_a - V_{c_i})^2}{2}\right) - \sum_{i=1}^{n} SO_i \cdot (\mu \pm tg\alpha_i + r_{ci}) = (M + m_r) \frac{dV_{ci}}{dt} \end{cases}$$

The above system of differential equations has n+1 equations, one for the air itself and n for the n containers(or trains) moving inside of the tube.

In the above system: λ is the Darcy-Weissbach coefficient; l is the lenght of the pipeline; d is the ID of the tube; ρ is the average airdensity; V_a is the airspeed in the tube; k is a coefficient associated with the dynamic pressure(for sucction propulsion k=0, for pressure propulsion k=1); CD_s is the pressure loss coefficient associated with the manifolds for pro½ulsion system; CD_i is the pressure drop coefficient associated with the entrance in the tube; $CD_c(sc_i)$ is the coefficient of the pressure drop associated with the container(which is dependent of the space(sc_i) position of that particular container(or train); $\pm$ sign is dependent of the fact that the container could go up the ramp or descend(that is could be pneumatically propelled or braked).

In the above system, $-AV_a^2+BV_a+C$ is the expression of the flowrate -pressure characteristics of a centrifugal machine; as is seen the equation for air in the above system is equalled with the pressure reserve of the propulsion machine($f(V_a)$) -for exemplification, the centrifugal machine caracteristics was used (today over 90% of the industrial installations(Pipeline Conveyors) use centrifugal fans/blowers or compressors).

The above system could be solved numerically or analitically. However the analytical solution involves Ricatti differentical equations very difficult to be used practically.

Qualitatively speaking, the laws of motion for a pneumatically propelled container in a tube with constant air speed could be described by the following relations:

$$a_c = \frac{\alpha^2 e^{\alpha t}}{a(e^{\alpha t} \pm ct_1)^2} \qquad V_c = -\frac{\alpha}{2a} \frac{(e^{\alpha t} + ct_1)}{(e^{\alpha t} \pm ct_1)} \qquad S_c = -\frac{1}{2a} \frac{\ln(e^{\alpha t} - ct_1)^2}{e^{\alpha t}} + ct_2$$

(Marcu 1978)

Where: t is the time; a_c is the acceleration of the container; V_c is the speed of the container; S_c is the space of the container ; the rest being constants depending of the form of the container and the rolling characteristics.

It is obvious that the laws of motion for a pneumatically propelled vehicle are exponential.

In case when the air is considered a compressible fluid the isothermal flow with friction could be used. This was proven correct in many cases in pipeline flow and also in turbine labirinth flow where large gradients in pressure and temperature are encountered. Generally the constant air flow could be described in isothermal conditions as follows:

$$P \cdot V_a = ct$$

Where: P is the static pressure of the air in the given section and V_a is the air velocity in that section. From thia relation it is seen that when we have a lot of pressure we have a corresponding low speed for the same massic flow rate of air.

The most precise description of the propulsion phenomena inside of the pipeline is the method of characteristics developed by Riemann before 1900 and used extensively by the shock tube and balistic people for the description of the transient flow inside of tubes.

The equations of continuity, momentum, energy form a system with hyperbolic differential equations of second degree(order) which are solved along the characteristic directions in the x,t space. In this case, x being the axis of the pieline and t being the time. The said characteristic directions being coincident with the directions of propagation for the sound waves in the said space x,t.

In principle the real phenomena are as follows: at time $t=t_1$ we have a set of parameters associated with the container in the tube-like airspeeds, pressures, densities, etc; at another time t_2 these parameters will change. The solutions for these problems are numeric. In a numeric step, physically speaking, because of the local conditions(like a sudden ramp or a descent) the container is changing the speed and automatically changes the local airspeed. Then at t_2 a wave is radiated from the front of the container which is characterized by the travelling speed c(speed of sound) and by the difference of pressure across the wave front. This could be described enough precise by the Jukovsky formula:

$$\Delta P_{wf} \cong \pm \rho_F \cdot c \cdot (V_{aF_2} - V_{aF_1})$$

Where: ΔP_{wf} is the difference of pressure across the wave front;

$\pm$ is depending if it is a sucction wave or a pressure Wave; ρ_F is the density of the air in front of the container(an average in time in this case); V_{aF2} is the airspeed at the end of the numeric interval, that is after a Δt(ussually cca 0.005 sec); V_{aF1} is the airspeed at the beggining of the numeric interval.

Consequently in the back of the container we have:

$$\Delta P_{WB} \cong \pm \rho_B \, c \, (V_{aB1} - V_{AB2})$$

that is another wave radiated from the back of the container travelling with the speed of sound(the notations are similar except B is denoting the back of the container).

The best illustration of these sonic phenomena is when the container is getting out via a flap valve and is blocking the propulsion manifold(Maneciu Prahova Class).In this case we have a sudden variation of the air speed from the cruising phase (when the flap valve is closed) to the exit phase(when the flap valve is opened).This is simply heard at the other end of the pipeline as a loud bang and as a sudden variation of pressure in the pipeline.For a 1.4 km 40" OD pipeline and around 10 m/s airspeed, the pressure shock travelling with the speed of sound is about 400 mmWG.

So the real physics of the pneumatic wheeled propulsion involves a lot of waves generated all the time by anything moving inside of the pipeline, and these waves travel with the speed of sound back and forth, transmporting the variations of pressure and flow, reflecting and attentuation meantime.The process is awesomefrom computational point of view,but however proved correct for the precise simulation and understanding of steam, air rocket launchers or shock tube research or even low speed container propulsion .

Recent advances in RISC chips,paralel computing and optic circuits and the corresponding increase in computer memory, made possible the aeromechanic simulation using Characteristics/Riemann method.

However simpler, economic methods could be used,reaching almost the same precision,because the fluctuations of pressure and flow are very gradual and the masses of containers are huge compared with that of the air in the pipeline and the wave effect could be neglected if the inertial term for the air(the mass of air) is taken into account.This was already proven correct by the water hammer computations made by the hydraulic engineers. Indeed these variations of pressure and flow travel with the speed of sound but taking into account the mass of the fluid inside of the pipeline , the results are almost the same.

Southern Ontario Technologies Inc.has access to resources for computer simulation of the aeromechanic and economic behaviour of these systems using either the incompressible,isothermal, polytropic or characteristics/Riemann approaches (including power consumption computations).

The Pipeline Conveyors Aeromechanics is another example of beneficial effect of the progresses made in some other areas:gas flow in the long pipelines,the limits of the isothermal flow,

,sonic waves and the method of characteristics,finite element methods.The dramatic advances made in computer technology and associated software made possible a vast array of simulation and visualisation techniques which will have also a big impact on marketing and research tools.

State of the Art,Hystorical Aspects,Terminology

The idea of the tubular transport is not new.The pneumatic post on small tubes(the cartridge dispatch system) was used for long times since early Victorian era till today.The first big installations were built around 1865-1870 in UK and USA(cross tube dimension:40"-50").The railway development and competition,the lack of mass produced tubes,lack of inexpensive electricity produced in large centralized power plants, made impossible for this new technology to take off(the tubes for the first large tube installation built at Battersea(UK) were made of cast iron sealed with molten lead).

The advent of the large tube technology(for gas and oil and water transport) as well as the electrification made possible the revival of the large tubular conveyors.

The revival was initiated in the sixties in West Germany(Deutsche Post ZentralAMT, R.Kirchheim,SEL and others).Almost simulatneously around 1969 USA,USSR,Romania,UK and around 1976-1980 Japan, by pioneers like Carstens,Vandersteel,Alexandrov, Coanda-Teodorescu,Baker,Hark,Nakamura,Kosugi,Tomita and organizations like:Transco,Tubexpress,Transprogress,INCREST,ICTCM,BHRA ,BTG,Sumitomo,Nippon,Daifuku, embarked on crash programmes centered on large tubular wheeled container transport.

The first Canadian large diameter capsular pneumatic Pipeline Conveyor(36" OD,crushed rock) was built in May 1989 in Stoney Creek Ontario by Southern Ontario Technologies Inc(Bowron/Hammant /Harvey Industrial Tires Group), and could be seen in operation now.

The introduction of this technology first time in Ontario and South Eastern Canada was made by Mike Marcu in June 1984(Kilborn Engineers-W.Reiter/P.Davis Group),October 1984(McMaster Univ.-Dept.Engineering),1985(Inco,TubeAir,Standard Aggregates,APAO group),1987(Thorne Pneumatic Systems group).

In Western Canada this technology was introduced by Tubexpress (Pembina,Polloway group) and EMEC Trading Co(SKB Transprogress / Bumerang group(cca 1984).

Despite some initial skeptic reaction from the mandarins of pipeline slurry and liquid capsule transport concept,the Canadian crushed rock industry received with enthusiasm this simple technology.It is a lot of confidence now in US and Canada that these Pipeline Conveyors will cut the transportation costs substantially and will bring into the economic circuit some difficult and abandoned areas like Canadian Rockies or the Canadian North.

There is a controversy about the denomination of these conveyors.The initial term of Capsule Pipeline was rejected by the crushed rock people as being innapropiate for objects weighing 5-10 tonnes and trains totalling up to 100 tonnes.The term accepted by mining and construction bulk handling sector and by companies like Southern Ontario Technologies is "Pipeline Conveyors or Large Diameter Pneumatic Transport with Wheeled Containers.The Capsule term was inspired by laboratory work on very small installations.

The Oil Crisis,the Environmental Crisis and the economic problems associated with the resource industries and transportation in general pressed for better solutions for moving bulk materials.

Because of their structural simplicity,the large pneumatic wheeled conveyors were considered as a natuxral class of economic alternatives for the dedicated bulk transport.

There are now over 60 industrial installations all over the World, the largest being Lilo 2 of 50 km close to Tbilisi-Georgia USSR(gravel and sand).USSR has now about 60 installations (many of them 24" ID).Romania has 3 and Japan 3.The total number of industrial installations is close to 70 and together with the pilot plants and demonstrator units they are reaching over 100.

Aside from the original industrial pioneer countries like USSR, Romaina,Japan , a lot of attention is paid to this new transportation technology in countries like: USA,Canada,South Africa,UK,W.Germany and Australia.The consensum of the specialists at the Freight Pipelines Conferences(Philadelphia-Pennsylvania 1985,Columbia-Missouri 1989) is that these systems could replace and have to replace succesfully in many cases the belt conveyor ,the slurry pipeline, the truck and the rail transport.

Way of operation-structure

The Pipeline Conveyors could be monotubular,bitubular or multitubular.The tubular guideway used for these conveyors could be made of on the shelf steel(Stelpipe), plastic(Dupont) or concrete pipelines(Concrete Pipe).Inside of these pipelines the wheeled containers move by air pressure or air sucction.These containers are made like long gondolas having at the ends wheels. The Pipeline Conveyors could be with rails inside of the tube and in this case the wheels are made of steel.Also in case when there are no rails, then the containers are rolling on rubber or polyurethane wheels directly on the inner tube surface.It is possible to use aircushion also or combination of the above.

There are at each end of the said container gondola, rubber gaskets to seal the air flow and to facilitate the pneumatic propulsion.However it is not necessary to have a permanent sealing contact- the gap between the gaskets and the tube could be around 1/2- 1/4 "(the labirinth effect).This gap is sufficient to assure the pneumatic propulsion of trains up to 200 tonnes and more.When a train of containers is used, the labirinth sealing effect of successive seals is almost equivalent with the case when the gap is zero and the container is like a sealed piston.

The flowrate could be organized in form of trains of containers or individual containers.

For propulsion ,fans,blowers or compressors could be used.Air pressure for moving the wheeled containers is ussually between 1-10 psi(in most cases up to 3 psi).The typical propulsion unit is: 2 psi,18000 cfm,200 Hp(for 2-3 km-2000 tph).The airspeed inside of the tube is around 10-15 m/s and the wheeled containers are moving at a slightly smaller speed.

The necessary power is in the range of 100HP/Km for 1000-2000 tph for rubber/polyurethane wheels and about 50 HP/Km for steel wheels rolling on steel rails.Of course the power is increased when the inclination of the route is increased.

There is no limit to the inclination of the pipeline because the containers are in pure rolling situation(they don't depend on adherence).Southern Ontario Technologies Inc specializing in large diameter tubular transport,is offering a class of mining cableless hoisting equipment based on tubular pneumatic propulsion where a 70 t load could be raised on vertical in a 3 m tube with only 15 psi pressure.

The Pipeline Conveyors could easily negociate curves of 100 m and in some cases down to 40 m.So they can cope with any terrain configuration making bends around the corner of 90° if necessary ,going up and down depending of the specific application(open pit,mountain,industrial crowded environement).

The loading and unloading installations are based on on the shelf subsystems used by the mining and construction industries over the years. The loading/unloading process could be made either static or dynamic.Southern Ontario Technologies Inc. is offering a wide class of dymamic loading/unloading subsystems for the Pipeline Conveyors based on On-the shelf units(vibratory/apron feeders,etc).

The progress made in the field of solid state electronics generated very reliable and inexpensive control systems based on PLC,intelligent sensors, machine vision,microprocessor based servosystems.The on the shelf control and communication systems used for railway marshalling yard, factory automation or material handling are succesufully transferred/used by Pipeline Conveyors.Some essential advances in thyristor based converters made possible very economic alternatives for linear motor propulsion compared with the very expensive classic variant based on power transistors.Southern Ontario Technologies Inc. is offering to industry handling and propulsion units based on LIM technology eliminating some cumbersome old systems like chaindrives.

Summarizing the above, the Pipeline Conveyors are the result of the natural evolution of the pipeline, rolling(wheel) and control technologies.They were able to incorporate the substantial advances like the mass produced steel, concrete or plastic large tubes, the air handling machines like the centrifugal or positive displacement blowers,compressors or fans,also advanced polymers for wheels, a vast array of solid state control systems already tested in other areas, also the heavy duty high capacity loading/unloading systems for the unit trains and so on.All these were tested over the years and could be easily combined in a reliable inexpensive Pipeline Conveyor.

Economics-Advantages-Costs

The costs mentioned below are reffering to the North American conditions but the economic and functional/environmental advantages of Pipeline Conveyors are even bigger in other areas because of their inherent qualities of "appropiate workshop technology."

The investment cost of these systems is in the range of about 1,000,000 $/km for the bitubular 24" system(flowrate capacity for 24" system: up to 4,000,000 tonnes/yr).For the bitubular 40" system the cost could be around 2,000,000 $/km.Southern Ontario Technologies Inc is offering to industry 24",36",40" systems -bitubular or monotubular for 1,000,000-2,000,000 $/km depending of the conditions of application.The cost of these systems is made of 80% pipeline guideway and about 20% the loading,unloading, containers,propulsion and automation equipment .

A propulsion blower/fan completely equipped(flow control,noise control,manifolds ,valving,etc) is about 50,000$.A container for the 40 " pipeline able to carry up to 10 tonnes is about 8,000 $(wheel assembly,gondola,gaskets,etc).The steel pipeline itself is in the range of about 900$/tonne.An interesting note is about the plastic pipeline which is totally corrosion proof. The pipelines supplied by Dupont Canada have almost the same price as the steel ones but of course other structural characteristics.Southern Ontario Technologies is offering to industry now both steel and plastic Pipeline Conveyors as well as concrete pipeline Pipeline Conveyors when the cost and conditions are favourable for concrete use.The plastic pipe systems could be used for sulphuric acid sludges,potash,salt and other corrosive materials.

Compared with other systems,the investment for the Pipeline Conveyors is usually half of that for the other systems of transportation for the same flow rate and distance.For example a belt conveyor for 1000 t/h is about 3,000,000 $/km and the pipeline conveyor system could be fabricated and installed for only about 1,500,000-2,000,000 $/km.In combination with the portable crushers the tubular Pipeline Conveyors could replace the heavy haulers in the open pit mines.

Recovery of the invested capital

The pipeline wheeled conveyor has a proven inflation-proof quality.This because it uses electrical energy for driving the fans or blowers and is fully automated.The inflation for these systems is below 1% essentially being the same --as the other pipeline transport systems(currently being around 0.6%).

It is possible to recover the invested capital either in short term or long term,that is either in 4 or 10 years depending on the financial conditions associated with the project.

Economic studies and economic simulation as well as the practical experience on these systems proved that it is possible to recover the invested capital in cca 4 years and still to be competitive with the already amortized existing systems(belt,rail, truck).So it is possible to include in the cost of transportation the cost of the invested capital and to recover it in about 4 years. After the "recovery" period,the cost of transportation in the case of Pipeline Conveyors decrease dramatically (essentially made only of energy and maintenance).

Operation costs

The operation costs for the tubular wheeled conveyors is as an average below 0.1 $/tkm.The cost is depending on flowrate and other factors.This cost could decrease in some applications depending on distance and flowrate as low as 0.01 $/tkm.Only a detailed feasibility study and a computer simulation for a specific application could indicate the probable cost for Pipeline Conveyor operation.

These systems have very few moving objects/namely the wheels and the motors/rotors of the blowers.Consequently they have a very low maintenance.Southern Ontario Technologies Inc., developed systems able to be virtually maintenance free(systems with heavy duty rails inside of the tube and steel wheels able to resist in free rolling conditions over 50 years).

The advantages of the Pipeline Conveyor Technology

The Pipeline Conveyors are structurally simple and the most important advantages of them are the following:

1.Totally non polluting,environmentally superior,noiseless,eventually unseen(buried pipeline).
2.Able to carry large flowrates on any distance in isolated, remote areas day by day and night with a high degree of safety.
3.Able to be installed in cities or crowded industrial areas(because they can take bends,go almost vertical and be burried if necessary).
4.Simple to design,fabricate,implement(appropiate workshop technology),based on classic on the shelf low technology and materials.
5.Totally independent of weather/terrain conditions(rain,water, snow,lizzard,hurricane,flood,avalanche,mountain,swamp,underwater,etc);could be underground,above the ground,in the air, under water.
6.Easy to integrate and use the on the shelf verified systems and products such as pipelines,wheels,solid state intelligent electronics,linear motor systems,etc.
7.Less expensive,essentially simpler than belt,truck,rail systems in term of operation and investment costs.Movable inside of the open pit(to cope with the blasting/excavation front.
8.Inflation proof(less than 1% per year).
9.Don't need maintenance road like the belt conveyors,they could transport multicargo in both directions,they don't have distance limitation,the maintenance could be made without stopping the system and could be designed to be easily dismantled and relocated.

Areas of application

The Pipeline Conveyors could be applied for:crushed rock transport,gravel and sand operations,ore transport,agricultural products,garbage transport,hazardous,toxic,radioactive,corrosive materials transport.Also they could be applied for transporting packaged goods or even people.

Granulation

The granulation of the material to be transported could be 300 mm for the 40" tube system or 200 mm for the 30 " tube system.With the advent of the portable crushers,the 100 mm granulation close to the extraction or blasting area is very common and the Pipeline Conveyors could be easily used.

Distances

There is no distance limitation for the Pipeline Conveyors. Systems of up to 50-100 km could easily built and operated.Open pit distances like 1,2,3 km are very attractive for these systems.Even the inplant variants like 200 m stackers or process conveyors could be implemented ecnomically.

The technical knowledge in this field permit today projects of 1ooo km or more(bitubular systems 24",40").

In an initial phase the Pipeline Conveyors could be used in Open Pit operations, from Pit/Mine to the processing plant or to the port/railway terminal without problems on distances from less than half of a mile to over 100 miles.

Conclusion

The above advantages make the Pipeline Conveyors a very attractive economic alternative for the dedicated heavy duty bulk transport.They have all ingredients in terms of simplicity,versatility ,safety and economics to have the confidence that the application of these systems could generate substantial savings and open new possibilities in the field of mining and construction transportation.

REFERENCES

1.M.I.Marcu,Short Note on the Romanian Tube Transport Program, Journal of Pipelines,6(1987)217-219.
2.M.I.Marcu,Tube Transport Systems at Expo'86,Vancouver,BC,Journal of Pipelines,6(1987)217-219.
3.M.I.Marcu et al.,Pneumocapsule Pipelines:Potential for North America, Journal of Pipelines, 6(1987) 221-238.
4.M.Rukavina,Air-Powered Conveyor Can Transport Crushed Stone, Rock Products,Vol 93 No 3 1990 pg 17 .
5.Pipeline Conveyors,Coal , Vol 26 No 8 p 82.

PIPELINE CONVEYOR PROTOTYPE

36" steel pipeline for Pipeline Conveyor.Propulsion unit in the far plane

Plastic, steel pipeline used for Pipeline Conveyor Technology -36 " diameter

Propulsion unit-centrifugal fan for Pipeline Conveyor

Propulsion unit-inlet silencer, fan,motor and control panel for Pipeline Conveyor 36 "

Propulsion unit-Pipeline Conveyor Prototype- exhaust silencer manifold and electric motor

Details pro½ulsion unit- inlet silencer, exhaust silencer,manifolds-vacuum, pressure,partial view of the fan

36" Pipeline Conveyor-the emerging sequence of a container at the end of the tube.Angle between wheels:120 degrees.

36" YUKON class container 8-10 tonnes capacity-gondola and gaskets-Pipeline Conveyor

36" YUKON class container -gaskets detail, wheels,gondola-Pipeline Conveyor

Front view of the 36" YUKON class container-detail on the front gasket(CD factor cca 2000(mmWG,metric, m/s))

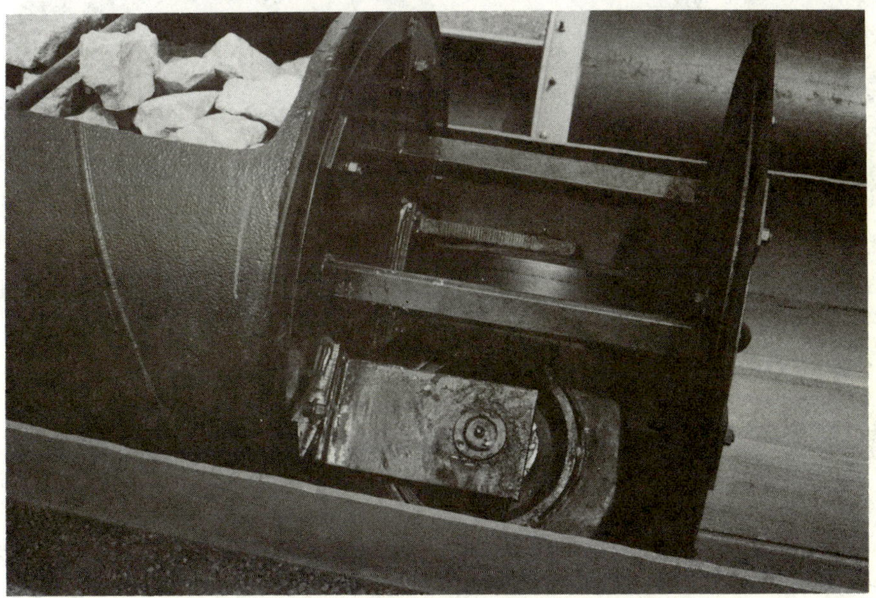

Wheel assembly detail - 36" YUKON class container-16" OD 5" wide-polyurethane-6000 lbf load per wheel

24" tubular system with rotary cluster arrangement;various other parts used in Pipeline Conveyor(wheels,shock absorbers, steel rollers).

24" container with a rotary cluster of wheels and gondola with absolute stability.The gondola body is resting/articulated on trunions on the said rotary cluster.

Coal tests proved that Pipeline Conveyor Technology is appropiate for coal transport

Crushed rock transport is one of the main application of the heavy duty Pipeline Conveyors

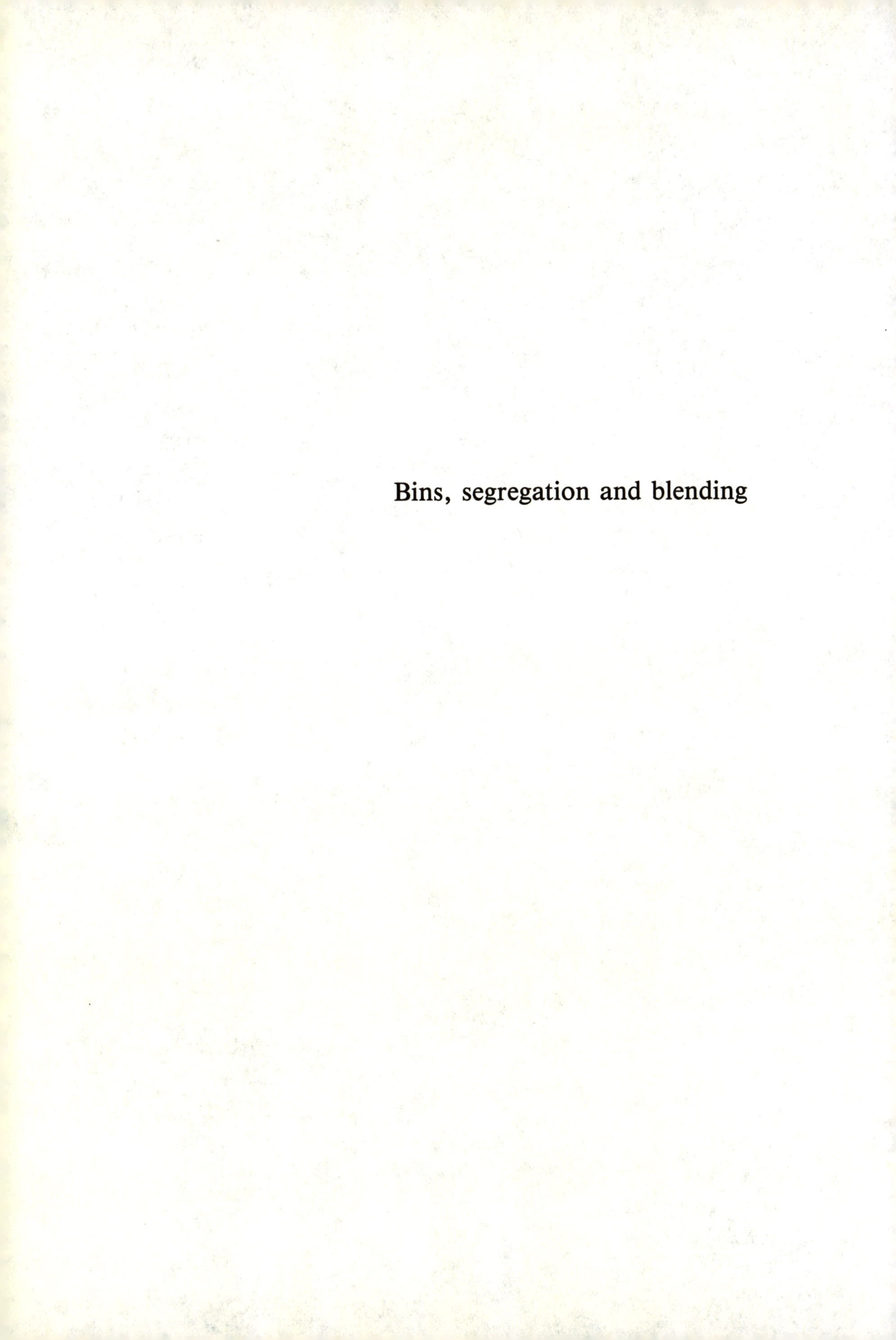

Bins, segregation and blending

Particle segregation upsets metallurgical processes

D.J. Goodwill and R.T. Jenkyn
Jenike & Johanson Ltd., Etobicoke, Ontario, Canada M9W 5X9

Many metallurgical processes are seriously affected by the effects of particle segregation. Mixtures of solid particles can separate or segregate while they are being handled. This often results in costly quality control problems due to waste of raw materials, lost production, increased maintenance and capital costs required to retrofit existing facilities. In pyro-metallurgical processes, segregation in the raw mix can lead to furnace explosions and serious injury to plant personnel.

In a plant manufacturing a chemical reagent, serious problems with an electric furnace were experienced as a result of particle segregation. Coke and lime were mixed in a 40/60 ratio and fed to the furnace via a surge bin. The materials were dry and free flowing. The top size of the lime was 75 mm and that of the coke 25 mm. The majority of the product less than 5 mm in size had been removed in both cases. Both materials undergo particle size segregation when centrally charged into a bin. The coarser particles roll to the outside of the bin, whereas the fines sift out and remain at and near the centre of the bin. The effect of this size segregation on the coke/lime ratio is illustrated in Figs. 1a and 1b. Samples were taken from the material discharge stream at regular intervals during the emptying of a pyramidal single outlet funnel flow bin used to smooth material flow to the furnace. This particle segregation had two very serious effects on the furnace operation. Firstly, high percentages of fines (up to 30%) led to decreased bed porosity and increased frequency and severity of explosions. Secondly, a mix deficient in either lime or coke reacts inefficiently and leads to greatly increased raw material usage.

In a plant manufacturing phosphorus, a mixture of phosphate rock, coke and silica was flowing down an inclined chute into a series of three bins, as shown in Fig. 2. Each of these bins fed a separate region of a furnace. The plant found that their quality control problems were the result of the mixture of ingredients not being uniform from one bin to another. In particular, they found that the first bin in the series had a high percentage of the fine silica particles, while the last bin in the series had a low percentage. Upon investigation, it was found that this was due to sifting segregation occurring as the bins were refilled. The mixture was placed at the top of the sloping chute section by a bucket elevator, and as long as there was any space left in the first bin or the vertical pipe above it, all of the material would enter this bin. As soon as this pipe became filled, the stream was automatically diverted to the second bin. However, as this process took place, some of the fines in the mixture sifted out as it passed over the vertical pipe of the first bin, thereby leading to a higher concentration of fines in the first bin

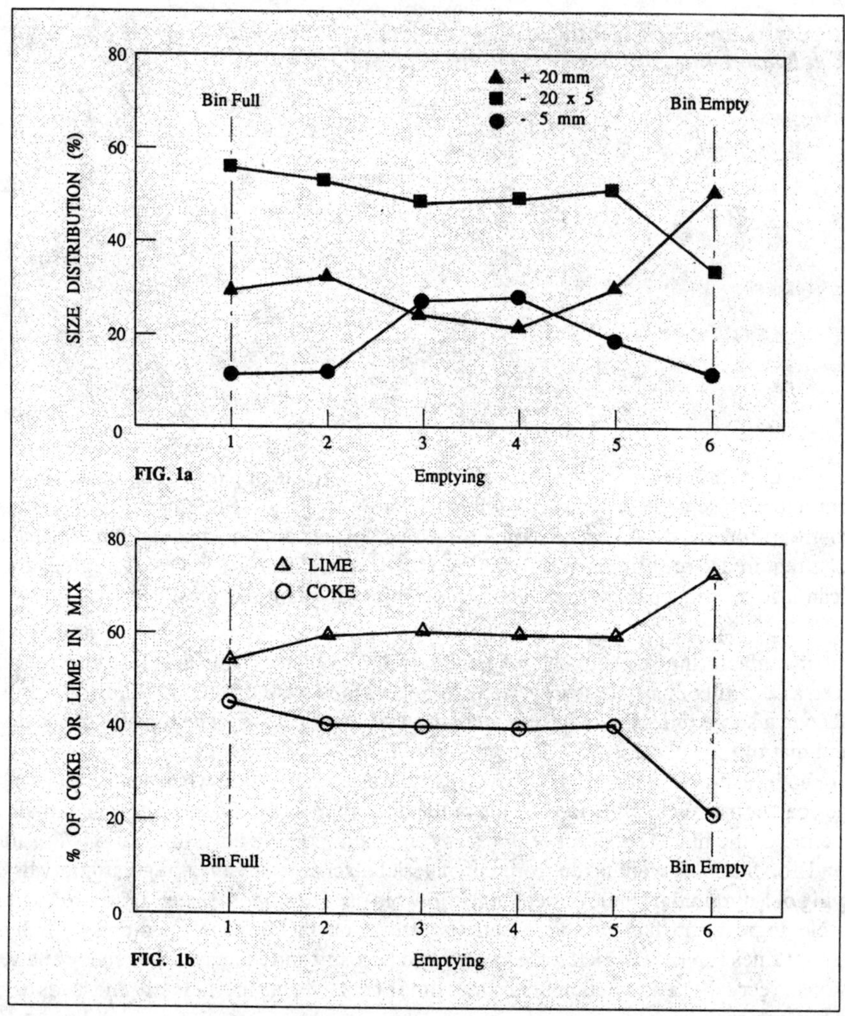

FIG. 1a & 1b. Particle size and product segregation in a funnel flow bin discharging a coke/lime mixture

than in the second. Later, as the second bin became filled and flow was diverted to the third bin, this process continued, with some fines being dropped out of the stream into the first bin, a smaller quantity being dropped into the second bin, and whatever was left over going into the third bin.

In a plant manufacturing metal strip by a process involving the cold roll compaction of metal powder, particle size segregation led to serious problems with the quality of the strip. Density fluctuations, strip separation and breakage caused a substantial production loss. The powder, which was produced by a process of atomizing and drying, was a mixture of spherical particles less than 30 mesh (600 micron) with about 30% passing a 325 mesh (45 micron) screen. Serious particle size segregation due

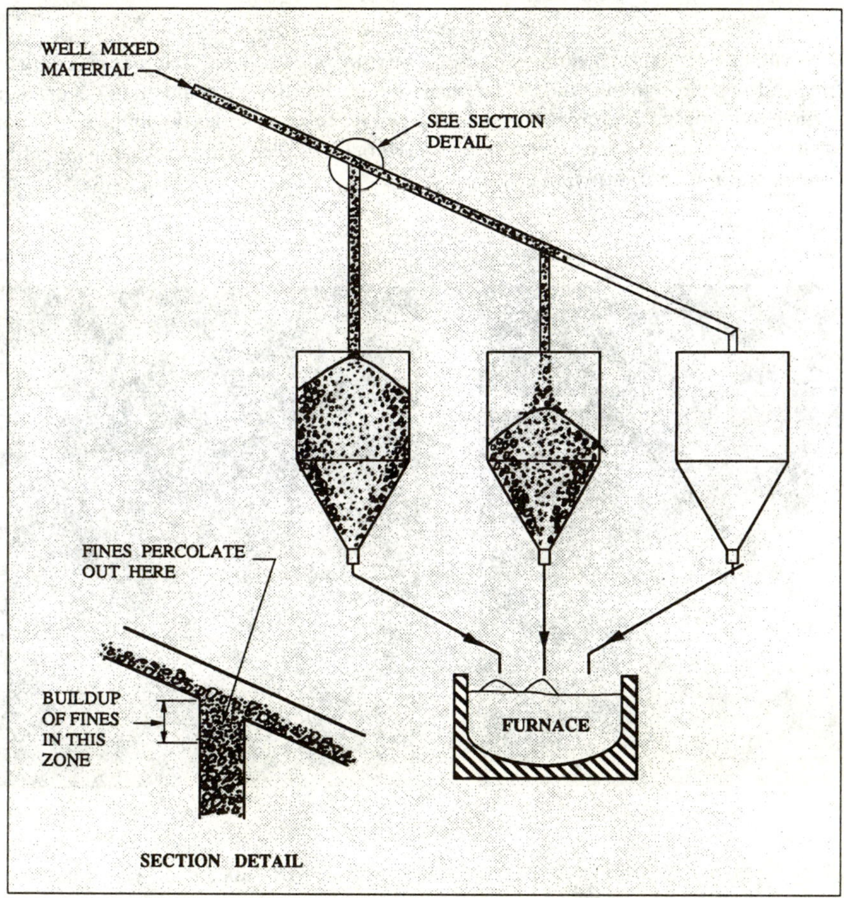

FIG. 2. Segregation due to sifting on a sloping chute

to sifting was occurring during discharge from a cone blender and from special hopper transfer cars used to feed material to the roll compactor.

In a plant manufacturing alumina, serious size segregation occurred during the unloading of a large silo into rail cars. The flat bottomed 20 m diameter silo was 20 m tall with a 2000 tonne capacity. The silo was loaded centrally at a single point and discharged through four openings simultaneously to a single rail car. Serious segregation was experienced. An average of 8% fines (minus 45 micron product) is acceptable in this product. However, rail cars frequently had fines concentration as high as 20%. This led to serious problems in the downstream process. In this case fine particles became airborne while the silo was being charged and secondary air currents carried them to the periphery of the tank where they were deposited in a narrow concentrated rim. As the level of the silo rose and fell, there was a continuous accumulation of fines at the periphery. On final discharge very high percentages of fines were measured.

Common Mechanisms of Segregation

The most prevalent segregation mechanism when a mixture of various particle sizes is being handled, is the sifting of fines through the voids of larger particles. For this mechanism to occur, there must be interparticle motion, such as horizontal shearing when a mixture slides down the surface of a pile. The fines sift into a stationary layer below and are left behind. The sifting mechanism is illustrated by the demonstration model shown in Fig. 3.

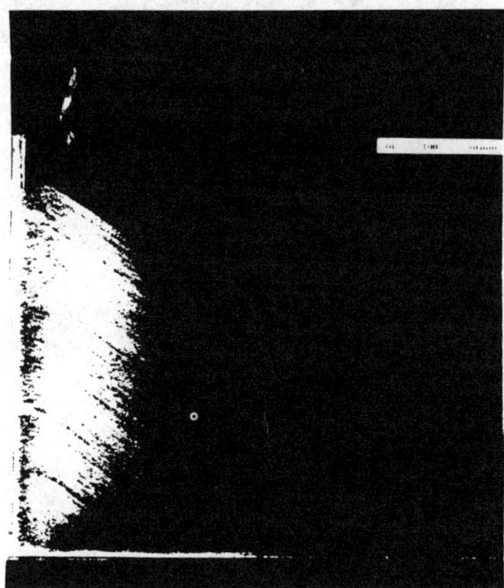

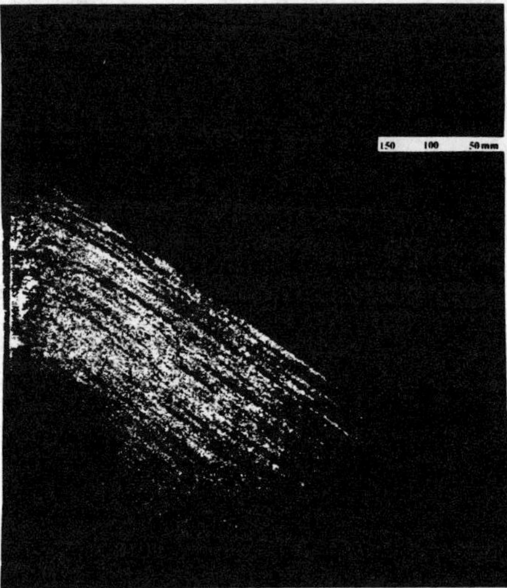

FIG. 3a & 3b. Sifting segregation during bin filling with a) 30:70 mix and b) 70:30 mix of corn grits and mustard seed.

Sifting occurs with very free flowing mixes in which there is a difference in particle sizes. For example, a binary granular system requires a size difference of as little as 1.5 to 1 for sifting to occur. Segregation effects are most pronounced when particles act independently of each other. Generally they do not occur with cohesive materials. For example, when a fine powder adheres to the surface of a coarse particle, no segregation takes place till the bond is broken. The tendency to segregate by sifting decreases proportionately as the mean particle size of a mixture reduces below 500 microns. Below 100 microns the tendency for segregation is usually quite low. The ratio of coarse to fine product is also important. The worst cases occur for mixtures containing more coarse particles than fines.

Fines can also percolate into the non-flowing boundary layer of a flow channel. Fig. 4 shows a photograph of a silo model bin after filling, partial emptying and refilling. The dark coloured material is fine, the light coloured is coarse. The percolation effect can be clearly seen.

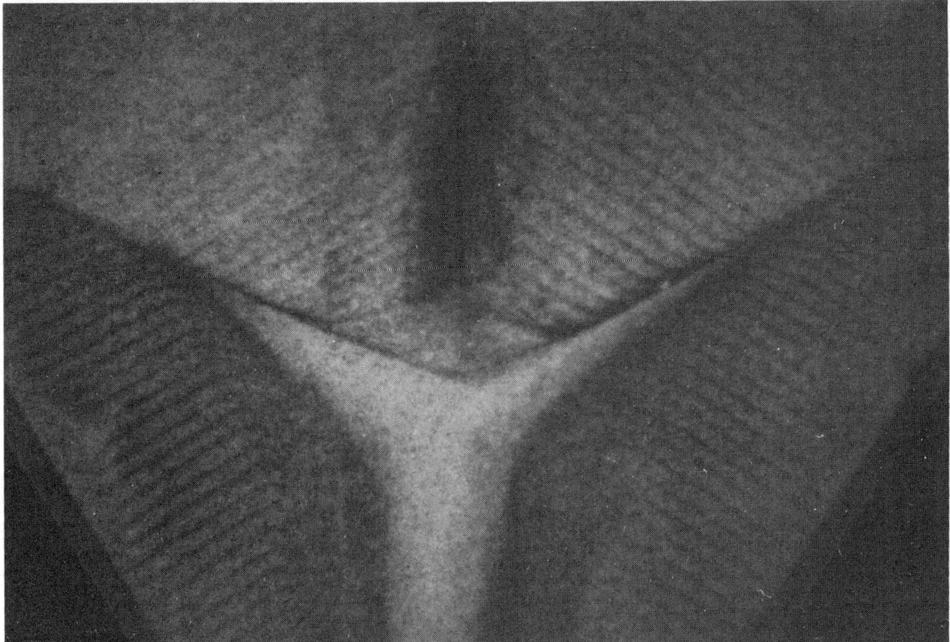

FIG. 4. Percolation of fines into a non-flowing zone.

Particle density differences can also lead to segregation effects. Fig. 5 shows a 50:50 mixture of similar sized particles of iron ore and calcite fed into a slice model. It can be seen that the darker coloured iron ore particles remain at the centre beneath the fill points whereas the lighter coloured calcite is displaced to the outside. Presumably particle density differences coupled with dynamic effects account for this type of segregation effect but large scale tests are needed to accurately measure the magnitude.

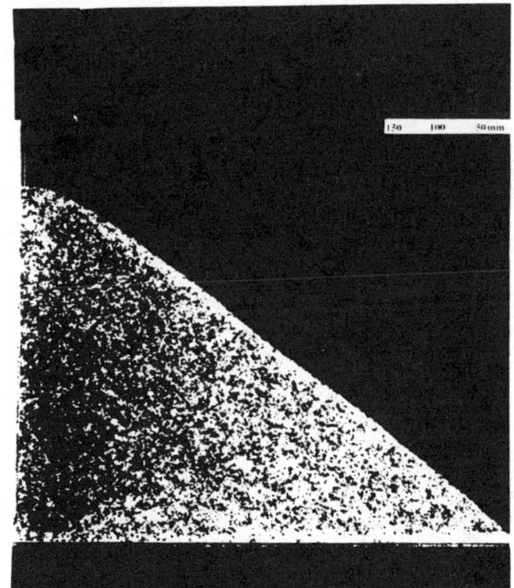

FIG. 5. Dynamic segregation during bin filling with a 50:50 mix of iron ore and calcite particles of similar size and shape.

Another common mechanism for segregation in bins is caused by very fine material separating from the bulk and becoming airborne during the bin filling. Particles smaller than 10 microns will form a dust cloud and continue to settle onto the top surface of the pile long after filling is completed. Larger particles, up to 40 micron for example, become airborne during charging into the bin and are carried by secondary air currents to areas of least disturbance at the bin periphery, or valleys between piles, where they rapidly settle and accumulate.

Goodwill and Carson (1) have described in detail other segregation mechanisms including differences in angle of repose, surface friction on chutes, fluidization of fines in bins, dynamic and trajectory effects. Williams (2) has carried out experiments to measure the segregation coefficients of 50:50 binary mixes of free flowing particles in terms of mean particle size and also the ratio of particle diameters.

Minimizing Segregation in Bins

Controlling the flow pattern in bins is the key to eliminating or reducing most segregation problems. In general, flow patterns can be classified as either funnel flow or mass flow. In funnel flow some material remains stationary while the rest is moving, whereas in mass flow all the material is in motion whenever any is withdrawn. Hopper configurations to cause mass flow usually have steep and smooth hopper walls. Typical flow patterns for mass flow and funnel flow bins are shown in Fig. 6.

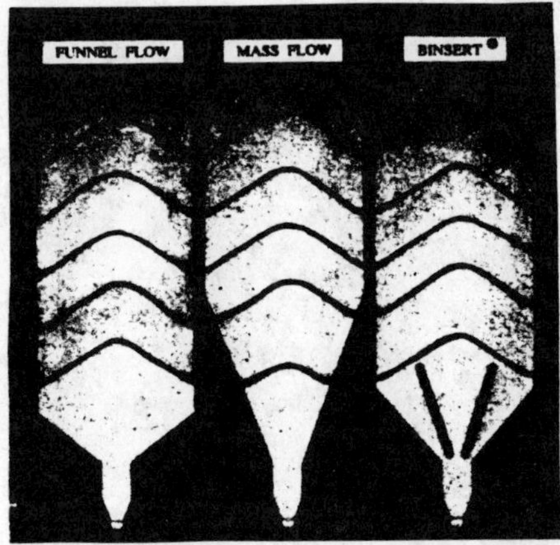

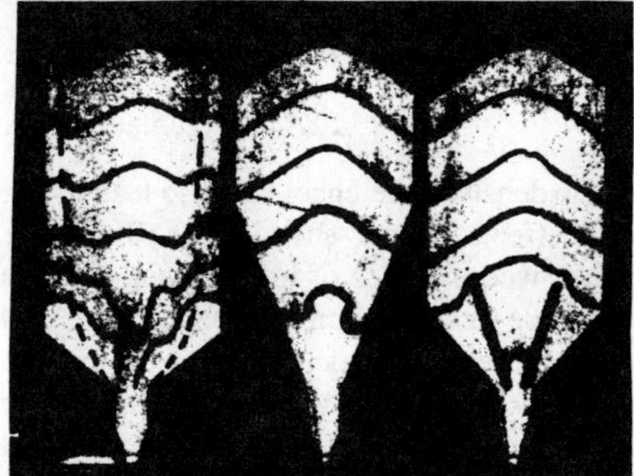

FIG 6. Model demonstration of flow patterns

Funnel flow occurs with shallow, rough walls. Most conical bins used in industry with hopper slopes of 45 to 60 degrees from the horizontal do not produce mass flow even with smooth liners. When a funnel flow bin serves as a surge bin, having a rising and falling solids level, it tends to discharge fine and coarse particles in alternative slugs at the same frequency as the fill cycle of the bin. This discharge cycle can be eliminated completely by designing the hopper to provide mass flow. In mass flow there is a uniform vertical velocity across the bin cross-section in the cylinder. This forces a horizontal cross section to be removed ensuring a remixing of particles that have segregated radially (Fig. 7).

To achieve mass flow, the hopper geometry and liner must be compatible with both the flow properties of the solid and the surface friction of the hopper wall material (3). The feeder must be designed to withdraw uniformly across the outlet and the solids in the bin must not fall below a critical level. This level varies with several factors, but typically the top of the pile should be at least 3/4 of a bin diameter above the converging hopper section.

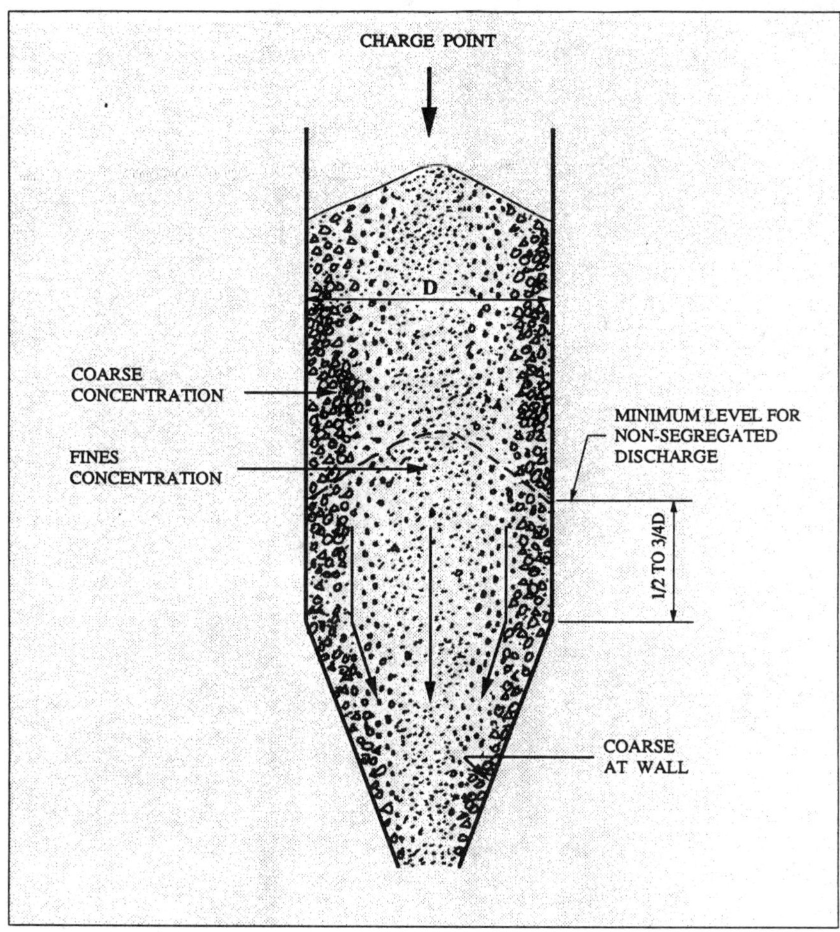

FIG. 7. Mass flow bin remixes particles which have segregated radially during filling.

A new concept has been developed in recent years by Jenike & Johanson for controlling flow patterns and converting funnel flow bins into mass flow bins. The BINSERT® system shown in Figs. 6 & 8 consists of a small hopper inside a large hopper. The geometry of the assembly is such that mass flow is produced inside the inner hopper and also in the annular space between the inner and outer hopper.

The BINSERT® has two main advantages over a typical mass flow bin. Firstly, the headroom may be substantially reduced. Secondly, whereas a mass flow bin requires a certain minimum material level to eliminate segregation, the BINSERT® can give perfect non-segregated discharge regardless of material level. The BINSERT® is covered by patents in North America and Europe (4).

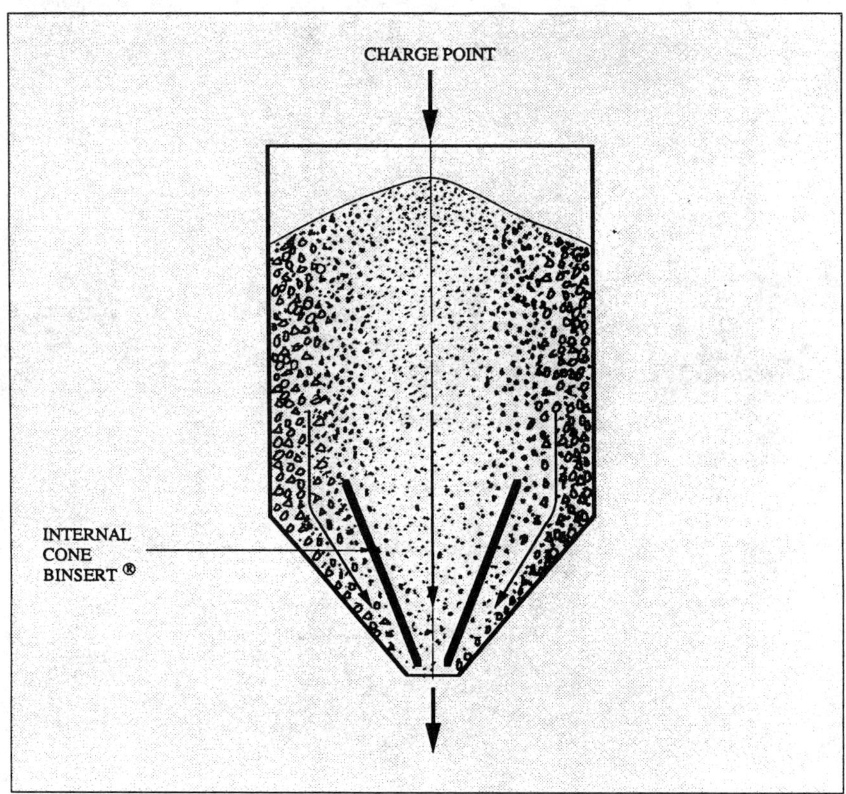

FIG. 8. BINSERT® can eliminate radial segregation.

References

1. Carson, J.W., Royal, T.A. and Goodwill, D.J. Understanding and Eliminating Particle Segregation Problems; Bulk Solids Handling Vol 6, No. 1, Feb. 1986.
2. Williams, J.C., and M.I. Kahn: The Mixing and Segregation of Particulate Solids of Different Particle Size; Chem. Eng., London, Vol. 19 (1973); p. 269.
3. Jenike, A.W.; Storage and Flow of Solids Bulletin 123., Utah Engineering Experiment Station.
4. Blending Apparatus for Bulk Solids. U.S. Patent 4,286,883, Sep. 1, 1981.

Vibration in materials handling systems

A.J. Matchett
Department of Chemical Engineering, Teesside Polytechnic, Middlesbrough Cleveland, TS1 3BA, U.K.

Synopsis

The British Materials Handling Board has established a working party to report on the effects of vibration in materials handling systems. There is a widespread ignorance of the implications of the use of these dynamic devices. A set of guidelines is being prepared as a source of information and advice to the user, including a questionnaire/checklist to be completed by equipment suppliers. The findings of the working party are reviewed, including sources of vibration, isolation and mounting, the effects of vibration on bulk solids and research. The effects of vibration are very complex and the understanding of fundamental principles is limited. However, if the installation and operation of such equipment is undertaken with care, then fault-free running can result.

Vibration in Materials Handling Equipment

Vibration is often purposely applied to bulk materials in order to induce flow in hoppers and silos, to maintain flow in vibratory conveyors, or for other purposes, such as screening and comminution. There have been reports of problems with such systems and, in an attempt to assess the nature and extent of these problems, the British Materials Handling Board organised a "face-to-face" meeting in June 1989.

The meeting brought together equipment manufacturers, users, contractors, academics (including the author) and representatives from industrial research establishments. The industrial representatives were first invited to present problems with vibrating systems which they had encountered. Typical examples of such problems include:

1. Whole buildings vibrating upon the installation of a vibrational device, due to inadequate damping and isolation. This type of problem appears to be more prevalent in newer buildings, which are of light construction and have relatively low inertia. Buildings also have resonant frequencies, which create great problems if they coincide with one of the harmonics of the equipment.
2. Support frameworks used for mounting vibrational equipment, which have shaken or failed because they were designed purely on the basis of static loads, or the equipment was mounted in unsuitable locations.
3. Failure of foundations on which vibrating devices were mounted, or other related problems due to the transmission of vibration through soils.
4. Two vibrating sources causing interference effects.
5. Transmission of vibration through the air by passageways acting like the pipes of an organ.
6. A vibratory feeder at the base of a hopper, which transmitted the vibration through the material into the hopper, causing the structure to

shake. The vibration was transmitted through the flowing material, as when the hopper was run empty, the hopper ceased to vibrate.

7. Severe structural damage to hopper outlets which discharge onto vibratory feeders.
8. Minor changes of geometry around vibratory feeder systems causing structural fatigue.
9. The use of vibratory flow inducers, which sometimes work very effectively, but in some circumstances do not, for no clear reason.
10. Materials adhering to vibratory feeder decks and hindering flow, inspite of the vibrational forces applied at the surfaces.
11. Flow problems at the outlets of hoppers onto vibratory feeders.

Members of the meeting stressed the need for careful mechanical design of vibrating equipment and the necessity for adequate damping and isolation. However, it was felt that there was widespread ignorance of the dynamic nature of vibration in industry as a whole. The destructive effects of shock loads and resonance were often underestimated or completely ignored. Equipment manufacturers pointed out that isolation of vibration had a cost associated with it. A reputable firm tendering for a contract and allowing for adequate isolation in the quotation could easily be undercut on price by a disreputable competitor who just sold a piece of equipment and left the customer blissfully unaware of the potential problems associated with the operation of that equipment.

It was pointed out that vibrating equipment would give few operational problems, if properly specified and installed. Furthermore, modifications can not be made on a piecemeal basis and any changes in operation or loading must be carefully considered and the manufacturer should be consulted wherever possible.

The effects of vibration are usually seen as a positive influence in promoting flow in bulk materials. However, vibration can also be very bad for such systems, causing consolidation and compaction, resulting in severe blockages.

The British Materials Handling Board responded to the situation by setting up a working party to draft a set of guidelines for vibratory equipment. The objectives of these guidelines would be:

1. To make users aware of the essential requirements for the installation and operation of vibrational equipment and make them aware of potential problems.
2. To present methods of solving such problems by damping and isolation and provide a statement on good working practice.
3. To review the current state of knowledge of vibrational research in materials handling.
4. To provide a check list for buyers of equipment, including relevant data they should seek from manufacturers and suppliers.

A brief review of some of the working party's findings is given in the remainder of this article.

Sources of Vibration

Sources of vibration can be categorised in two major groups:

1. Deliberate vibration.
2. Unplanned vibration.

Deliberate vibration is the result of the planned use of vibrational equipment such as vibratory flow inducers, vibratory feeders, screens, certain types of mill and polishing and deburring equipment.

Unplanned vibration results from external equipment or the environment, including out of balance motors on rotary equipment, compressors, pumps, mills and mixers. Environmental sources include road and rail traffic with transmission through the ground and the air.

It may be necessary to protect hoppers, chutes and other equipment from vibration because of the detrimental effects they can have on flow properties.

Isolation of Vibration, Mounting and Structural Consideration

If the sources of vibration are not isolated in some way, then the vibrational energy will be transmitted. Isolation can be achieved by either counterbalancing or damping.

Damping is the result of mounting the vibration source on a suitable non-transmissive medium. This may take the form of a massive block with very high inertial forces, or a non-elastic material which will absorb and dissipate energy but not transmit it.

Counterbalancing occurs when the vibration source is mounted on springs, on a counterbalance base which is also mounted on springs. The spring stiffness and mass of the counterbalance weight must be carefully chosen so that the majority of the vibration energy is converted into motion in the counterbalance weight.

Vibrational devices are dynamic and this must be taken into account in the design of mounting and related structural work. The magnitude and direction of the dynamic forces that the equipment applies to its mounting is constantly changing and is quite different from the static loads. This may lead to fatigue, if the mountings are not properly designed. Furthermore, shock loads can occur at start-up and shut-down. Correct design and installation procedures will reduce these effects to acceptable levels and provide a good basis for trouble-free operation.

Vibration Questionnaire

The questionnaire provides a checklist of important information to be completed by equipment suppliers. It emphasises the dynamic nature of such equipment and the information sought includes:

- The magnitude and direction of dynamic forces.
- The mode of vibration and type of vibration inducer.
- Frequency, amplitude and direction of vibration.
- Methods of isolation and their efficiency.
- Recommendations for mounting and positioning.

Completion of the questionnaire by all competing suppliers will enable the customer to make a fair comparison of all quotations and see to what extent he is comparing like with like.

The Effects of Vibration on Particulate Solids

The effects of vibration upon solids are extremely complex and depend upon many factors such as particle size, density, packing, system geometry and the amplitude, frequency and direction of the applied vibration. Under certain circumstances, vibration causes bulk solids to dilate and under other circumstances, vibration causes the compaction of beds of material. The simplest manifestation of compaction is the tapping test giving the Hausner ratio [1][2]. Roberts developed a modified Jenike cell which could be vibrated and was able to demonstrate the apparently contradictory finding that if vibration was applied before a shear test then the sample contracted and a higher shear stress was required to initiate motion. However, if vibration was applied during shear, the sample dilated and a lower shear stress was required to maintain motion. Therefore, vibration can be beneficial to bulk flow, but can also have a detrimental effect.

These observations show that it is necessary to consider the stages through which material passes as it is subjected to vibration, including:

1. The condition of the material before the onset of vibration-particle size distribution, moisture content, bulk density and packing, restraining stresses, geometry.
2. The onset of vibration-inertial effects, out of balance forces, shock loads.
3. The conditions under sustained vibration-amplitude, frequency, direction of vibration, bulk density, particle motion, segregation, contraction, dilation, interparticle contact forces, fluidisation. This stage of vibration is often, wrongly, considered as pseudo steady state.
4. The cessation of vibration-inertial effects, over-run, shock loads.
5. Conditions when vibration has ceased-dearation, settlement, changes in packing and density.

Results show that if conditions are measured in stage 3 above, they will be different to stage 5 and this explains, in part, the observations of Roberts.

Several attempts have been made to measure friction in vibrating systems [3][4][5]. As a result of these experiments, it is sometimes stated that vibration reduces the coefficient of friction. Vibration reduces (and can also increase) interparticle forces and if such results are analysed in terms of actual normal forces, then it could equally well be argued that while forces have been reduced the coefficient of friction has increased. However, it must be remembered that in many cases, forces are only reduced for half of a vibration cycle and are increased in the other half of the cycle as the direction of acceleration is reversed. The implications of this increase in forces over part of the vibration cycle have been largely ignored.

Vibratory conveyors have been analysed and modelled [6][7]. Models deal with point masses and take no account of internal circulation and other effects, but even in these simple models there are very complex flow regimes, involving forward, reverse and static cases.

It may be concluded that the effects of vibration upon particulate solids are not well-understood and have sometimes been misinterpreted. Vibration can aid flow, but may also hinder flow and there is a tendancy to concentrate upon the positive aspects and ignore the problems. This lack of fundamental knowledge is also reflected in industrial practice, where there is a great (and sometimes very successful) reliance upon empiricism. However, the use of a vibrating device is not a universal panancea for all flow problems. Roberts [3] reports pilot plant tests on a hopper with damp sand, where vibration was seen to improve flow in some situations, but had no effect in others, with no apparent explanation as to why.

Current Research

There are problems inherent in this area of work. The major one is that of time-scale. There are two conflicting time-scales at work in materials handling systems. At the lower level there is the scale at which things can visibly be seen to happen in hoppers, chutes or related equipment. Flow processes usually occur in a scale of seconds or large fractions of seconds - the long timescale. However, below this there is vibrational acceleration which is changing cyclically many times a second - 50 to 60 Hz if mains frequency is used. In order to monitor changes at this level, then instrumentation is required which can capture data at a rate of hundreds or thousands of times per second -the short time scale. The capture of this data, its analysis and relation to the much slower time scale at which individual particles can be seen to move presents major problems as the short timescale is several orders of magnitude smaller than the long timescale.

Most previous work has only looked at events on the slower timescale and only recently have workers begun to investiage at the faster and more fundamental level. Many industrial processes work at this short timescale, including:

Vibratory conveyors
Vibratory flow inducers in hoppers.
Equipment which introduces shock loads.
Impact loads, in the filling and emptying of hoppers.
Accidental and Environment vibration sources.

A research group has recently been established at Teesside Polytechnic, Department of Chemical Engineering to investigate these areas. Current themes include the transmission of vibration through bulk solids and adhesion of particles onto vibrating surfaces. The work is jointly funded by SERC and British Steel. This work requires a different type of instrumentation to that usually used in powder testing. Signals from measurement devices must be sampled at a rate of the order of hundreds of times per second. This is very rapid compared to the speed of a conventional Jenike cell test. Furthermore, the wave form of the output signal needs to be presented and analysed-averaging the signal will destroy all the information at the short timescale. This requires load cells and distance transducers with very rapid response times. Likewise accelerometers and related devices must be used to measure the transmission of pressure and distortion waves. The signals from these devices must be passed to rapid Analogue-to Digital Converters, linked to fast PC's for data logging and analysis.

The modelling of processes at the short timescale presents problems of computation and interpretation. The response of a body resting upon a vibrating bed of powder of relatively simple failure and flow properties can produce very complex motion. Models have been developed of the response of incompressible Mohr-Coulomb solids, rigid-plastic solids and elastic-plastic solids in simple testing equipment.

It is intended that the experimental and theoretical developments will enable us to develop a better understanding of events at the short timescale and to apply this understanding to the design and operation of materials handling equipment.

References

1. H.H. Hausner, Powder Technology, 1981, 30, 3.
2. K. Kawakita, K.H. Ludde, Powder Technology, 1970-71, 4, 61.
3. A.W. Roberts, Chapter 6, "Handbook of Powder Science and Technology", Fayed and Otten (Eds), Reinhold, 1986.
4. T. Akiyama, T. Naito, T. Kano, Powder Technology, 1986, 45, 215.
5. T.P. Fisher, D.S. Coleman, Powder Metallurgy, 1974, 17, 302.
6. K. Erdesz, A. Szalay, Powder Technology, 1988, 55, 87.
7. K. Erdesz, J. Nemeth, Powder Technology, 1988, 55, 161.

In-line sampling and weighing

Advances in technology for complex custom feed material treatment at Noranda

M. Bédard, M. Chapados, and G. Kachaniwsky
Noranda Minerals Inc., Horne Division, Rouyn-Noranda, Quebec, Canada

ABSTRACT

The Noranda Continuous Smelting Process at the Horne Smelter was commissioned in March 1973. The ability of the process to handle a wide variety of feed materials and a steady stream of technical innovations in recent decades has served to maintain the Horne as an efficient and versatile custom smelter.

This paper reviews recent technological development and capabilities in areas such as sampling and preparation of secondary and complex materials, as well as the smelting and related environmental control issues. Future trends and improvements are also addressed.

---

(1)Production Manager

(2)Manager, Business and Technology

(3)Smelter Superintendent

INTRODUCTION

The Noranda Horne smelter commenced operations in 1927. It was originally designed to treat 910 tonnes per day of copper concentrates and ores coming from the new Horne mine site. In the 1930's, the company opened its doors to other mines in Northwestern Quebec. Treatment of feed materials on a custom basis has therefore always been an important factor in the operation of the Horne smelter. Throughout the years, the Horne gradually increased its custom smelting capacity, while maintaining its production from the Horne mine. By the early 1950's, 65 percent of the materials processed was from outside customers, and by 1974, this figure had increased to more than 85 percent. With the closure of the Horne mine in 1976, the company extended its smelting operations beyond ore and concentrate, and began to treat more scrap metals and residues rich in copper and precious metal content.

Corresponding with the ever-changing pattern of feed sources, the Horne smelter has adopted technological improvements to meet new and challenging business conditions. The technology of the original smelter of 1927 has been completely replaced over the years in the quest for improvement. One of these major changes occurred in 1973 with the start-up of the Noranda Process Reactor. This extremely efficient smelting system has been exploited to treat a wide variety of copper and precious metal (PM) containing custom materials. This paper reviews recent developments and improvements achieved at the Horne smelter, in the receiving, sampling and treatment of copper and PM scrap and secondaries, with particular emphasis on the development of new technology and equipment in complex feed materials treatment.

MAJOR SMELTER DEVELOPMENTS AND PRESENT OPERATIONS

The original Horne smelter in 1927 consisted of eight roasters, two coal-fired reverberatory furnaces, two Peirce-Smith converters, along with the necessary auxiliary equipment. Plant throughput was gradually increased and two additional roasters were built; the reverberatory furnaces were extended and converter diameters increased. By the 1950's, the proportion of custom feed had increased, typical roaster feed of this period being illustrated in Table I.

In 1957, a third reverberatory furnace based on wet-charge smelting was commissioned to handle additional receipts of custom material from Northern Quebec and Ontario. To cope with increasing matte tonnages, the converters were lengthened, and a new 4.27 m x 9.75 m unit, the largest converter in the world at that time, was commissioned in 1965(1). The anode plant was upgraded with the start-up of new rotary anode furnaces and the introduction of gaseous reduction of anode copper based on Noranda-developed technology.

While technological changes continued to be recorded at the plant in the nineteen sixties, a new era in smelting operations at the Horne began with the start-up of the Noranda Process on March 1st, 1973. The Noranda Process Reactor has been described in previous papers(2-7). The ability of the process to readily smelt a heterogenous mixed charge was recognized at an early stage of process development. The reactor has always handled a certain proportion of copper scrap, metallics, purchased secondaries and complex concentrates. This proportion has increased over the years, and typical smelter feed for 1988 is shown in Table I.

Major developments and changes at the Horne smelter since 1973 are summarized in Table II. As shown, modern equipment for handling, sampling and sample preparation of complex feed materials has been installed in the 1980's.

The present operations at the Horne consist of the Noranda Process, a reverberatory furnace with oxy-fuel burners, two oxygen plants with a total capacity of about 580 tonnes per day, five converters (two to three converters operating), three anode furnaces and two casting wheels. An acid plant with a 1,950 tonnes per day capacity is in operation since the end of 1989. A smelter flowsheet is shown in Figure 1.

Table I - Typical Horne Smelter Feed - 1950's and 1988

Material	1950's		1988	
	Feed Rate Tonnes/Day	Analysis Cu % (Average)	Feed Rate Tonnes/Day	Analysis Cu % (Average)
Noranda Horne Concentrate	340	11.6	0	-
Noranda Horne Ore	1,460	1.7	0	-
Noranda Group Concentrate	0	-	620	26
Custom Concentrate	620	21.0	1,520	23
Copper Scrap	0	-	195	35
Secondaries & Misc. Materials	30	20.0	185	25
Flux	310	-	420	0.2
Total Feed to Smelter	2,760		2,940	

On a daily basis, the individual feed components treated in the Noranda Process Horne smelter can vary widely. For example, feed can include conventional chalcopyrite-type concentrates, silver-bearing residues with little or no copper or sulphur, gold-bearing pyritic concentrates with no copper and mixed metallic scrap. The Noranda Process feed system and reactor are capable of comfortably handling feed sizes up to 100 mm and moisture levels up to about 14 percent. In all cases, the Noranda Process Reactor has proven itself to be versatile in handling a wide range of materials varying in both physical and chemical quality.

The smelting rate at the reactor has increased year after year. Figure 2 shows the net increase in tonnages treated over the years. The Noranda Process has not yet reached its limit in terms of variety of materials and tonnage treated.

Table II - Major Developments - Horne Smelter (1973 - 1991)

Year	Development
1973	- Noranda Process Reactor commissioned on March 1st, 1973 on direct copper production
1975	- Commence ladle cooling of Noranda Process slag
1976	- Rebuild one hot-charge furnace as wet-charge furnace and close five of the ten roasters - Horne mine closed
1977	- Close remaining roasters and last hot-charge reverberatory furnace
1980	- Installation of first converter waffle hood
1982	- 465 tonnes/day oxygen plant on-line and upgrade of Noranda Process feed system - Commence oxy-fuel firing on one reverberatory furnace and shutdown of second reverberatory furnace
1984	- Expanded facilities for scrap receiving and sampling, including the shredding system
1985	- Installation of two 1.3 tonne induction furnaces for scrap sample preparation
1986	- Shipment of lead concentrate commenced from Noranda Process cottrell to Brunswick Smelting at Belledune, New Brunswick, Canada
1988	- Commence construction of 1,950 tonnes/day capacity sulphuric acid plant - Installation of Noranda Sampler for high grade electronic scrap - Installation of Noranda Sampler for P.M. scrap
1989	- Partial closure of last remaining reverberatory furnace and commencement of Noranda Process as sole smelting vessel - Acid plant start-up - New equipment and facilities for scrap sampling and sample preparation and installation of two additional induction furnaces - Record Noranda Process production received in a calendar month period - 62,300 tonnes in March 1989
1990-1991	- Installation of a pebble mill for slag treatment - Replace the last reverberatory furnace by Peirce-Smith converter smelting technology

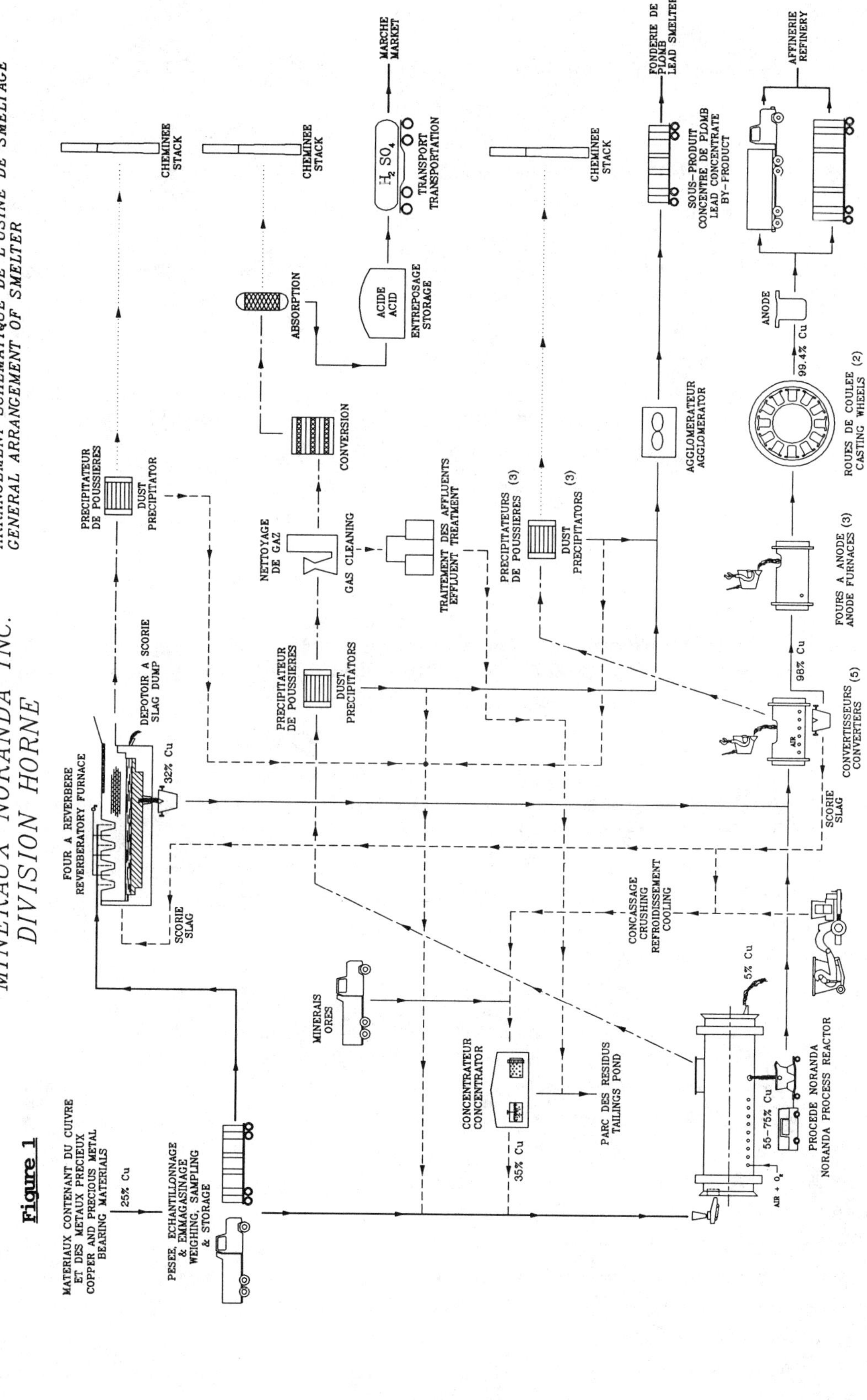

Figure 1

MINERAUX NORANDA INC.
DIVISION HORNE

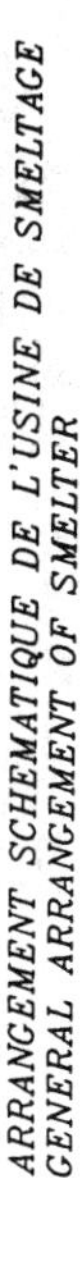

ARRANGEMENT SCHEMATIQUE DE L'USINE DE SMELTAGE
GENERAL ARRANGEMENT OF SMELTER

MATERIAUX CONTENANT DU CUIVRE ET DES METAUX PRECIEUX
COPPER AND PRECIOUS METAL BEARING MATERIALS
25% Cu

PESEE, ECHANTILLONNAGE & EMMAGASINAGE
WEIGHING, SAMPLING & STORAGE

FOUR A REVERBERE
REVERBERATORY FURNACE
32% Cu
SCORIE
SLAG
DEPOTOIR A SCORIE
SLAG DUMP

PRECIPITATEUR DE POUSSIERES
DUST PRECIPITATOR

CHEMINEE
STACK

MINERAIS
ORES

CONCENTRATEUR
CONCENTRATOR
35% Cu

PARC DES RESIDUS
TAILINGS POND

PRECIPITATEUR DE POUSSIERES
DUST PRECIPITATORS

NETTOYAGE DE GAZ
GAS CLEANING

TRAITEMENT DES AFFLUENTS
EFFLUENT TREATMENT

CONVERSION

ABSORPTION

CHEMINEE
STACK

ACIDE
ACID
ENTREPOSAGE
STORAGE

H_2SO_4
TRANSPORT
TRANSPORTATION

MARCHE
MARKET

CONCASSAGE
CRUSHING
REFROIDISSEMENT
COOLING

5% Cu
55-75% Cu
AIR + O_2
PROCEDE NORANDA
NORANDA PROCESS REACTOR

SCORIE
SLAG

AIR
CONVERTISSEURS (5)
CONVERTERS
98% Cu

PRECIPITATEURS DE POUSSIERES (3)
DUST PRECIPITATORS (3)

CHEMINEE
STACK

FOURS A ANODE (3)
ANODE FURNACES

AGGLOMERATEUR
AGGLOMERATOR

ROUES DE COULEE (2)
CASTING WHEELS
99.4% Cu

ANODE

SOUS-PRODUIT
CONCENTRE DE PLOMB
LEAD CONCENTRATE
BY-PRODUCT

FONDERIE DE PLOMB
LEAD SMELTER

AFFINERIE
REFINERY

Figure 2 - New Tonnage Treated in the Noranda Process (tonnes/day)

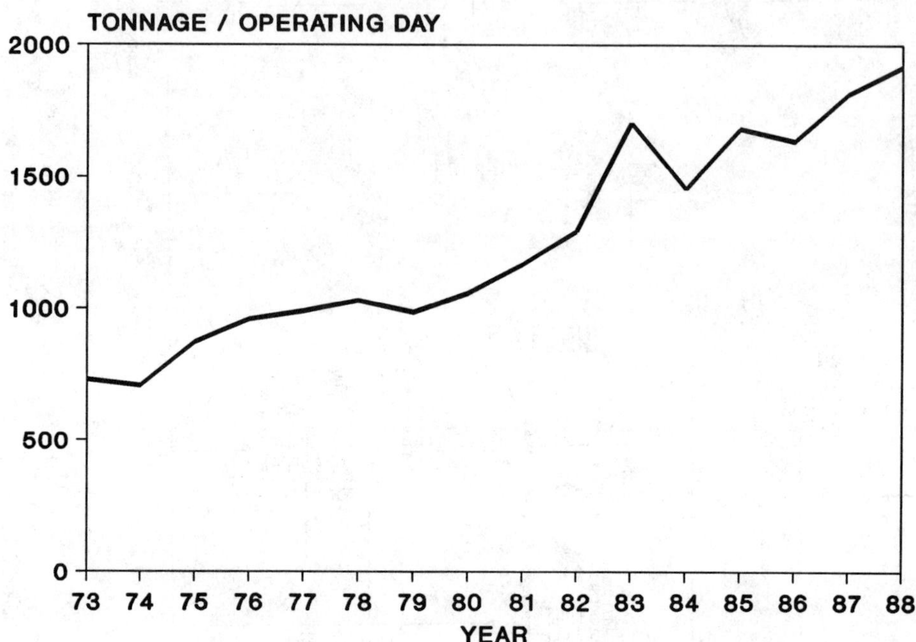

Copper and PM concentrates have always been handled, sampled and treated according to set procedures. With the increasing proportion of scrap and secondaries as feed materials, the Horne smelter has had to develop new procedures and install more sophisticated equipment. Over the last few years, important improvements in the handling, sampling and treatment of these materials have been achieved.

SCRAP AND SECONDARIES FEED TO THE SMELTER

Noranda purchases a variety of scrap and secondaries from suppliers across North America, Europe and Asia. Typical scrap processed at the Horne smelter includes slag, ashes, residues, wire cables, PM containing ingots, jewelry, telephone scrap, automobile parts and PM containing computer and electronic scrap.

Figure 3 shows the increased amounts of scrap and secondaries treated at the Horne smelter. From 23,000 tonnes per year in 1970, the tonnage has increased to 115,000 tonnes per year in 1988. This now represents approximately 15% of total feed to the smelter as illustrated in Figure 4.

Secondary materials are smelted in the reactor and in the converters. Because of increasing matte grades, opportunities to feed scrap to the converters have been substantially reduced. In order to treat the increasing quantities of scrap, a method of rendering more scrap suitable for treatment by the Noranda Process was necessary.

In 1984, a shredding plant was installed to pre-process scrap to maximize Noranda Process feed. Figure 5 shows the layout of the shredding circuit which reduces the scrap to a size of 2 or 4 inches (5 or 10 cm). Operating characteristics have been described in an earlier publication(8). The successful operation of the shredder has enabled the Horne smelter to significantly increase the total scrap tonnage treated. In 1988, approximately 50% of the scrap was shredded before being smelted in the Noranda Process.

Receiving

The scrap receiving department can accept both trucks and railcars, and unload the material, whether it be loose, in barrels, drums, bags or card-board boxes. All incoming material and samples are accurately weighed on stationary scales.

Material received in 1988 was typically 50% copper-bearing scrap, 25% slag and residues, 25% PM scrap and secondaries. Upon receipt, the material is unloaded, weighed, sampled and shredded if necessary, before being trucked to the reactor or the converters for smelting.

Depending upon the material classification as regards hazardous wastes, dangerous goods, etc..., the empty trucks or drums containing the material are cleaned at a decontamination station. All residual materials and washings are treated in the slag concentrate dewatering circuit from which slag concentrate containing solids from the decontamination is recycled to the smelting vessel. This efficient decontamination station has enabled the Horne smelter to accept a wide variety of feed, while ensuring all safety, health, environmental and transportation regulations are met.

Sampling

As a custom smelter, the Horne must determine the metal content of all scrap for settlement with its suppliers. Sampling of some PM scrap is performed at other Noranda owned companies, however, the Horne smelter is itself equipped to directly sample most varieties of copper and PM secondary materials.

Figure 3

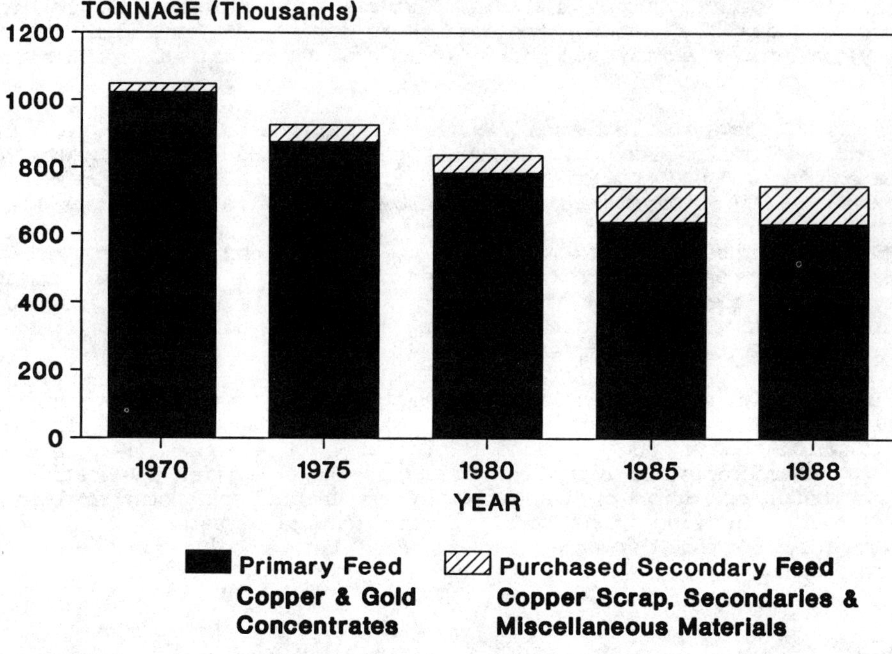

Figure 4

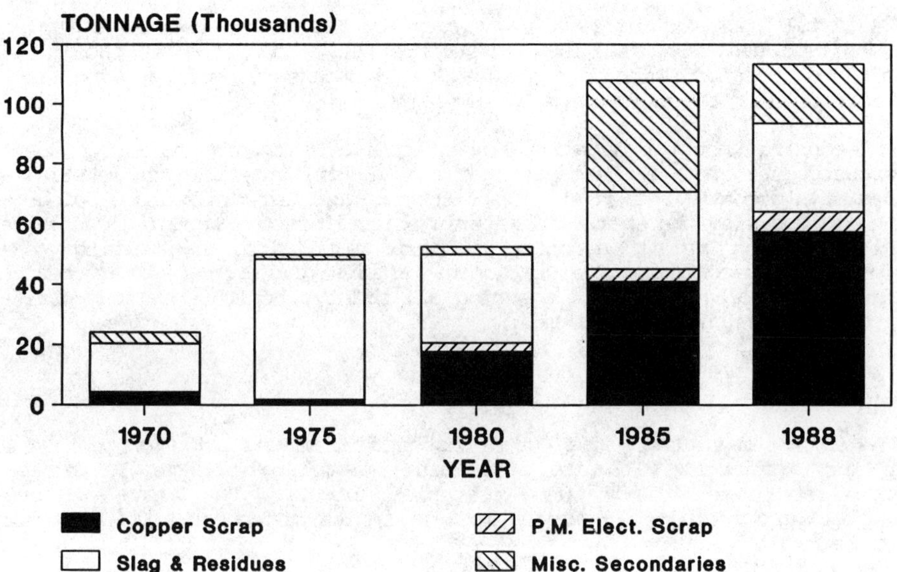

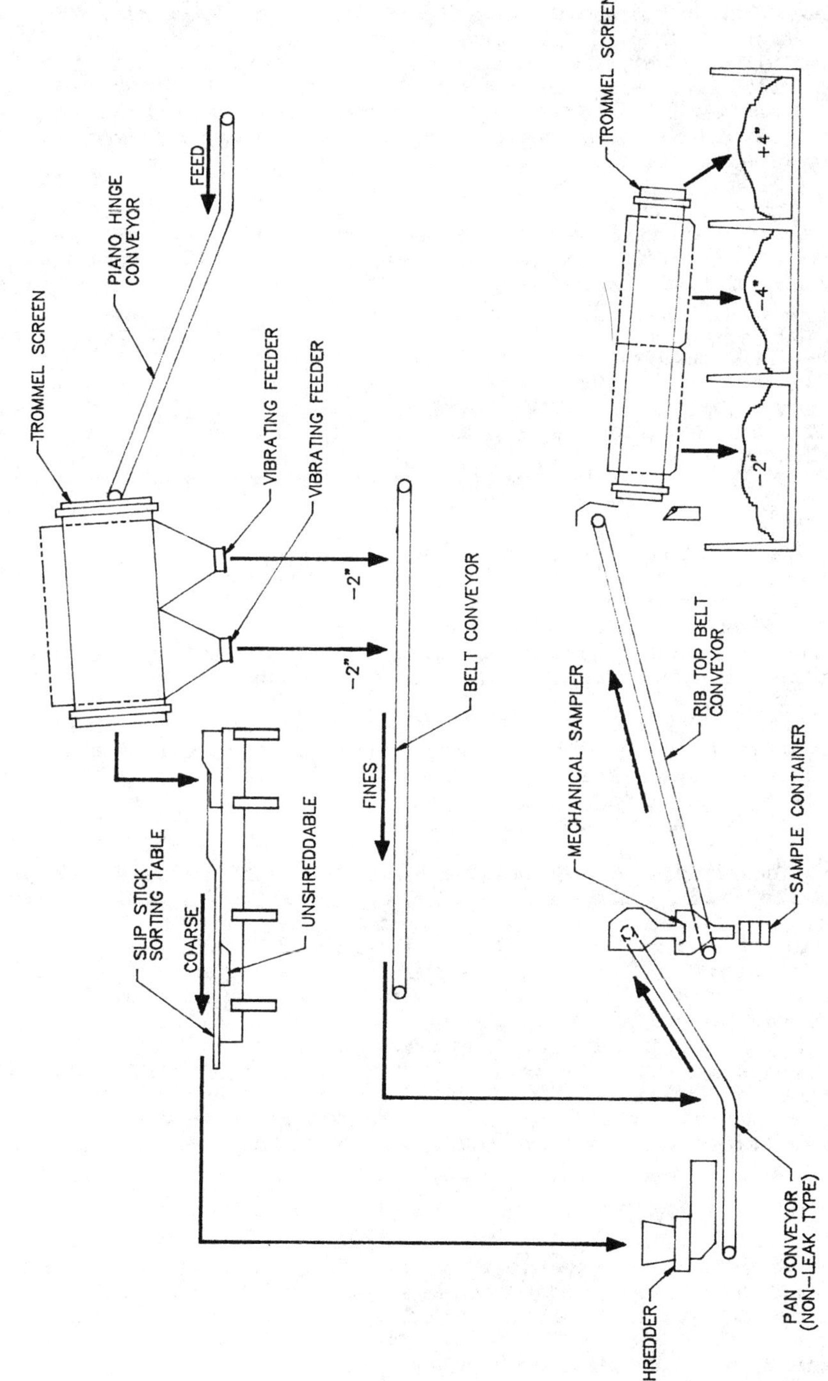

Figure 5

Shredding System Flowsheet

Several methods and procedures have been developed for obtaining representative primary samples from various types of scrap. The method used and also the size of the primary sample taken depends largely on PM grade, physical characteristics, chemical homogeneity and lot size of the scrap. In all cases, duplicate samples are taken and one sample is kept in reserve until final settlement with the supplier. The most common primary sampling methods employed at the Horne smelter have been previously published(9), and may be summarized as follows.

Copper concentrates, fine materials such as slags, crushed matte, and fine carbon or residues are pipe sampled. The pipe or the thief sampler is inserted into the drum or bag containers until it reaches the bottom. A minimum of five samples is taken. The aggregate weight must be at least 1 percent of the lot weight, and must total a minimum of 35 kg. Large shipments of coarse slag are processed through the mine crusher and sampled using an in-line Vezin sampler.

Homogeneous low grade ingots are sawed, and the ingot sawings are collected, bagged and sent to the laboratory for assaying. High grade ingots or ingots of poor quality are remelted in the induction furnace, and samples are taken by pin-tube sampling, which has been shown to be precise and accurate(10).

Grab sampling is used for sampling copper scrap (bales and sheets), motors, alternators, radiators and other automobile parts. The method consists of randomly hand-picking pieces to be representative of the entire lot for subsequent melting in the induction furnace. An automatic belt sampler is presently being installed at the discharge end of the shredder to replace the grab sampling practice for copper and low PM electronic scrap.

Until the end of 1988, PM electronic scrap was sampled using the increment sampling method. This consists of spreading the scrap onto a clean, flat metal sheet until a uniform maximum height of 15 cm is reached. Equi-spaced scoop samples are then taken. The number of scoop samples depends on the grade of the material being sampled.

The Noranda Sampler, Figure 6, is used for materials such as PC boards and electronics, chopped silver wires, diodes and computer chips. It has replaced the Horne increment sampling method. The Noranda Sampler was developed by the Horne smelter and was first installed at Noranda Sampling, Rhode Island, U.S. to replace the Gilson splitter. Following successful testing at Noranda Sampling(11), a sampler was installed at the Horne division and commissioned in January 1989. The design is based on the principle of taking a full-cut of a falling stream part of the time. It has 40 chutes positioned on the perimeter of a horizontal carousel which direct material to sample, reserve or bulk containers. The Noranda Sampler, as installed at the Horne, has the flexibility of selecting two inter-penetrating samples in increments of 2.5 percent. This apparatus is being patented by Noranda.

Sample Preparation

After a primary sample is taken, the particle size and sample mass are reduced to obtain a representative sample for laboratory assaying. The Horne employs several methods and procedures for preparing samples for assaying. These methods are outlined in the overall scrap sampling and sample preparation scheme shown in Figure 7.

Figure 6 - Noranda Sampler

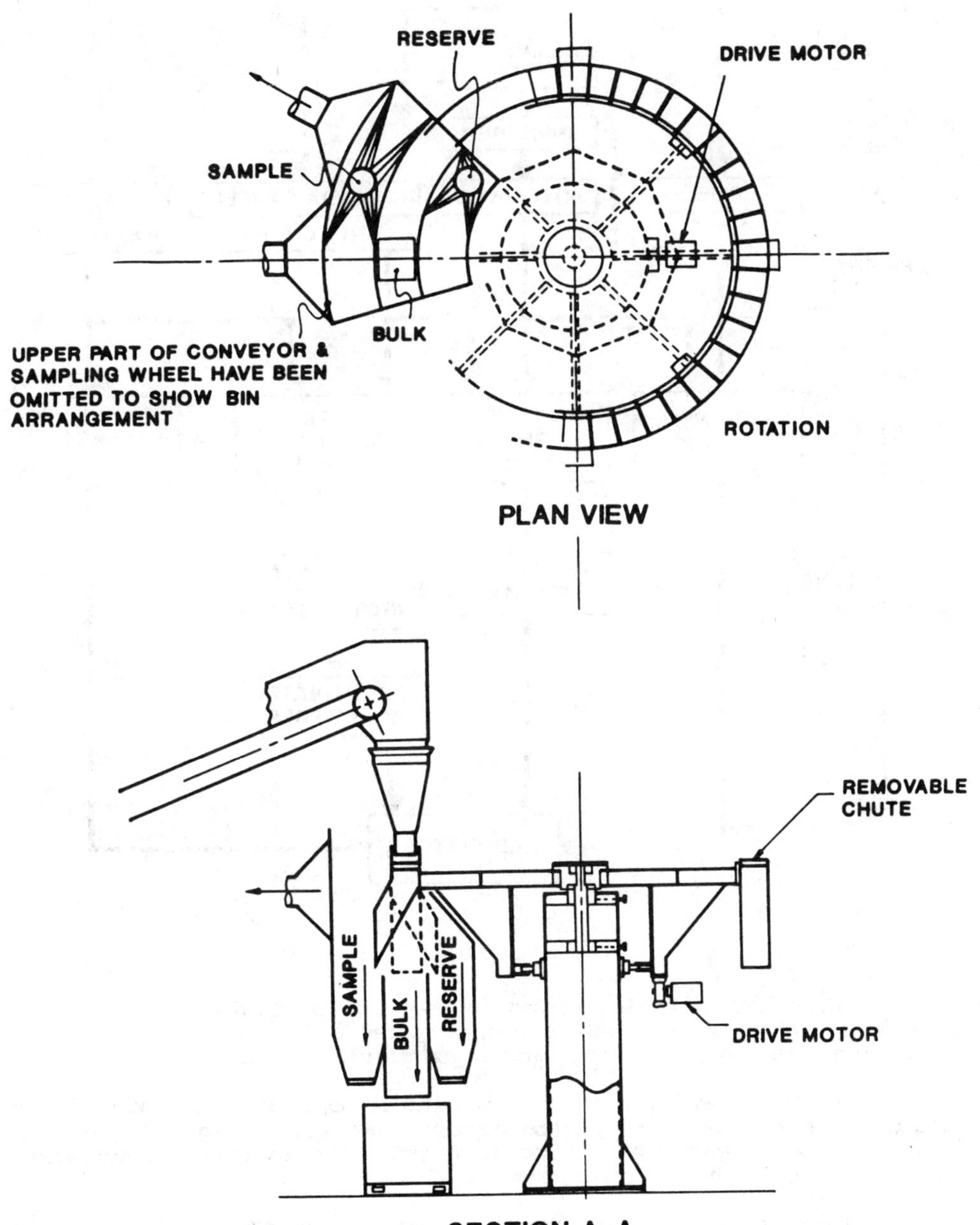

Figure 7 - Overall Scrap Sampling Scheme at Horne Division

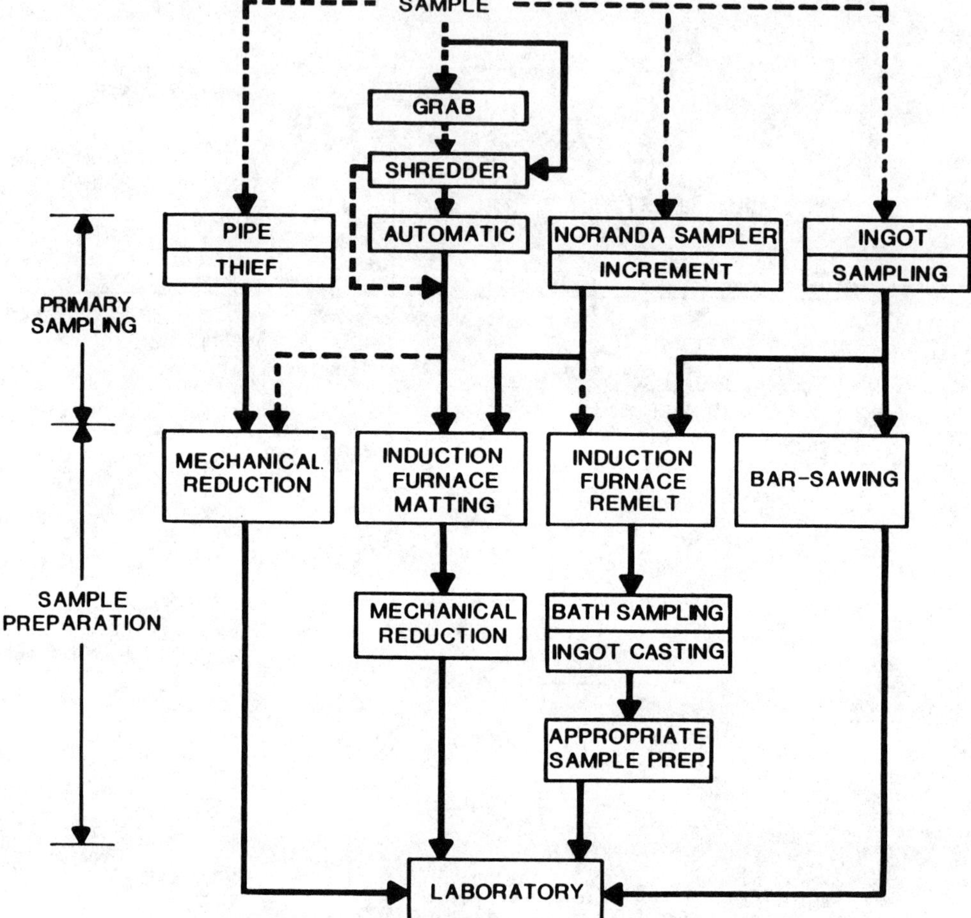

The mechanical reduction method is used for materials that are easily crushed to a fine powder such as ashes, mattes and slags, and consists of conventional crushing, screening, mixing and dividing procedures.

Bar-sawing is used for ingots of uniform shape, size and composition, containing less than 2,000 g/tonne Au and 7,000 g/tonne Ag. The bars or ingots are sawed according to set patterns. The sawings are collected, bagged and sent to the laboratory for assaying.

The induction furnace melt/remelt method is used for materials such as high PM ingots, bales of copper, motors, generators and radiators. This method was described in a previous publication(10).

For preparing samples of electronic scrap such as printed boards, the Horne matteing process is used. The process flowsheet is shown in Figure 8. A detailed description of the process has been given elsewhere(9).

The process consists of heating sub-lots of electronic scrap in a 1.3 tonne induction furnace. Pyrite is then added to the baked scrap and melted, leaving a molten bath of iron sulphide (FeS). The metals contained in the scrap (PM, Cu, Pb, Al, etc...) dissolve in the molten FeS forming a solution of metal sulphides.

The resultant matte is crushed to 6 mm in a jaw crusher, and then to -10 mesh in a pan muller, followed by screening. Any +10 mesh material that contains metallics is re-matted. The -10 mesh material is divided in a rotary divider to a subsample of 20 kg. The latter is then processed through a Gilson spinning riffler until a 2 kg subsample is obtained. This subsample is ground to -115 mesh in a ring mill and sent to the laboratory for assaying.

The matteing process reliability has been tested and compared to the conventional method of preparing electronic scrap samples. The conventional method consists of baking the sample to make it friable, followed by ball milling and screening to separate the oversized metallics from the fine powders (sweeps). The sweeps portion is then resampled, while the metallics portion is homogenized by melting with copper before being resampled.

The differences in Au assays between the matteing process and the conventional method for duplicate electronic scrap lots containing 300 gr per tonne Au or less, were found not to be statistically significant at the 95% confidence level.

The advantage of the matteing process over the conventional sample preparation method is that only one sample is obtained. In addition, less equip is used in the matteing process, material handling is easier and chances of cross-contamination are less. Tests have also shown that the FeS matte produced in the matteing process has a high capacity to digest aluminum, and can accommodate up to 20% aluminium, with a pyrite/PC board sample ratio of 2.

Analysis

Demand for better quality control, greater productivity and faster turn-around time for analysis has increased over the last few years. The Horne laboratory has turned to automation, computerization and intensive use of instrumental techniques. Inductively-coupled plasma (ICP) emission spectroscopy together with fire assay has proven to be a highly successful technique for the analysis of precious metal scrap samples(12). The recent introduction of Separate Sampling and Excitation Analysis (SSEA) coupled with ICP spectroscopy offers additional flexibility(13).

Figure 8 - Horne Matteing Process Flowsheet

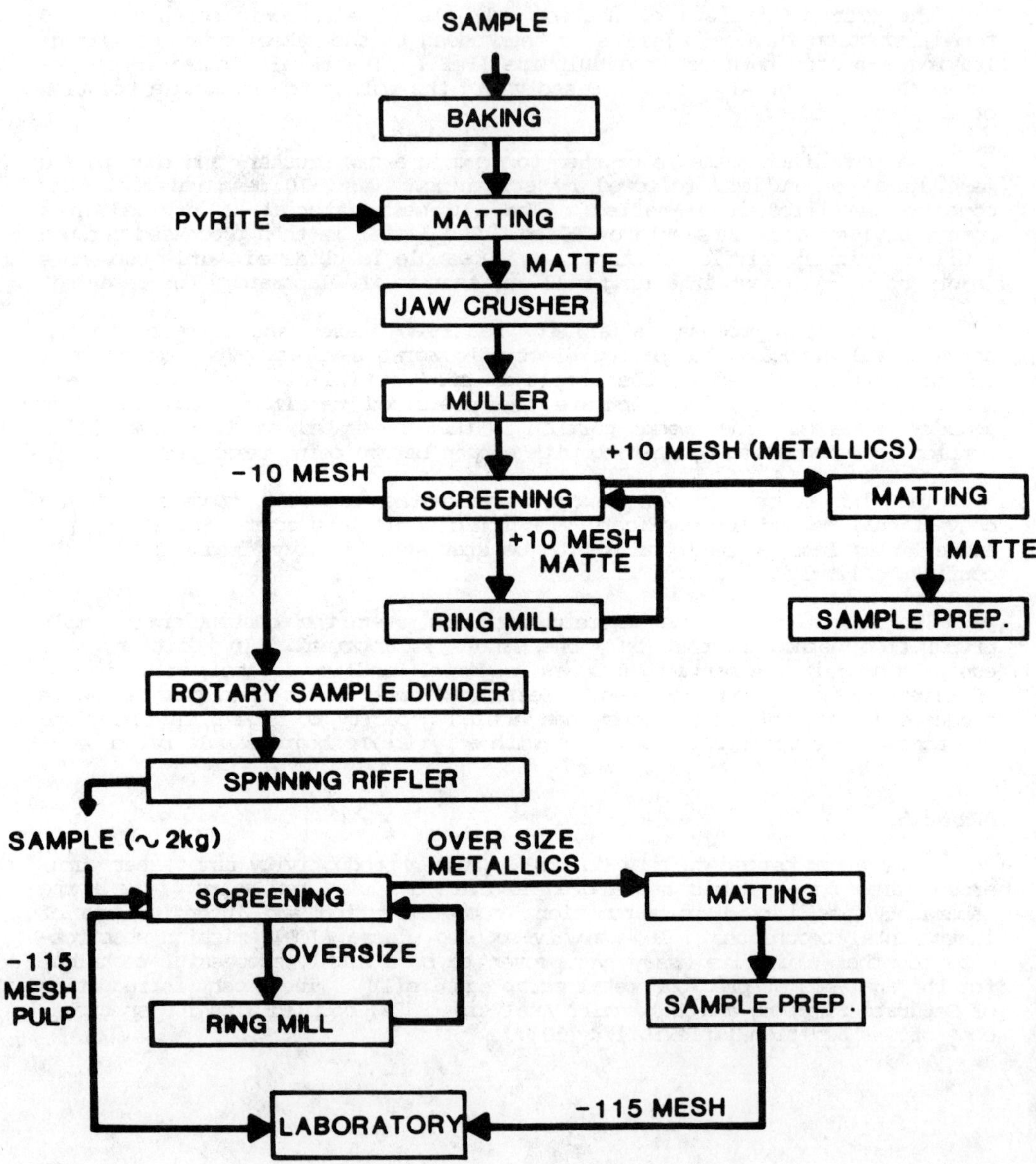

THE EFFECT OF SCRAP FEED ON THE SMELTER

Feeding

The wide variety of materials received is classified, shredded if required to a nominal size of -4 inches (10 cm) and stored in the reactor storage building in different areas based on several criteria: copper content, iron content, reactivity and on impurity or minor element content.

The scrap is then blended with conventional copper concentrates and fed through a 15 cm grizzly. Even reasonably fibrous scrap materials can be mixed in this fashion to provide acceptable Noranda Process feed. As long as the material can be handled on a conveyor system, and pass through chutes and transfer points, it can generally be treated.

The mix of scrap, copper concentrates, revert materials, fluxes and coal are stored in the Noranda Process feed bins.

Smelting

With Noranda Process instantaneous scrap treatment rates increasing in recent years to levels frequently exceeding 20% of the traditional feed, the effect of secondaries on Noranda Process metallurgy has been closely monitored.

Originally heat and mass balance calculations used for process control, ignored the effect of scrap on the process. Operating results were generally not unduly affected by this omission, and acceptable matte grade control, slag quality and operating temperature were maintained. Good feed blending, as described above, and the lower proportions of secondaries fed were the major cause of this.

Higher levels of secondary treatment indicated a need for a re-evaluation of this thesis, and the following secondary material parameters are now monitored in order to establish efficient Noranda Process operating control.

- The copper content and its effect on matte grade and quantity;

- The iron content and the flux requirements to slag off this iron;

- The effect of aluminum and refractory oxides on slag viscosity;

- The heat generated by scrap combustion and its effect on heat balance, or heat required by inert scrap. The use of the tuyere pyrometer(14) is also an invaluable tool in monitoring the Noranda Process heat balance;

- The overall oxygen demand per tonne of secondary material;

These parameters are presently being monitored by Horne Division technical staff. Results to date have shown that good control is possible when treating large and varied tonnages of secondary materials.

Off-Gases

The impact of secondaries on particulate loading of Noranda Process off-gases has been monitored and well controlled with high efficiency ESP's. Of more interest has been the impact of the organic content of scrap treated on the quality of Noranda Process off-gas.

Intensive testing has been conducted to determine if toxic organics, such as PCB's dioxins or furans, were present in Noranda Process off-gas and recycled dusts. These investigations were carried out while the Noranda Process treated unusually large quantities of organic, including PVC, containing secondaries. The testwork(15) did not detect the presence of these compounds.

It is felt that the extremely intensive smelting conditions in the reactor; the highly oxidizing, turbulent, high temperature conditions, and relatively long residence time, all contribute to the total destruction of these compounds.

Further measurements will be conducted at lower detection levels. This program will be undertaken as part of the future acid plant operating checks.

As noted above, a 1,950 tonnes per day acid plant, based on Chemetics' technology, handles the entire flow of gases from the Noranda Process. Considering the complex nature of the materials treated in the Noranda Process Reactor, particular emphasis has been placed on designing an efficient and versatile gas cleaning section of the acid plant. Included in this part of the acid plant is a conditioning tower, a Venturi scrubber, ten electro-static mist precipitators, ten shell and tube coolers and two Norzink mercury removal towers. The contact section of the acid plant is based on a simple three-pass, single absorption design.

FUTURE CONSIDERATIONS

Although Noranda Minerals is actively engaged in a major exploration program, the present trend is for copper concentrate feed sources to decrease in the north eastern region of Canada. Treatment of complex materials provides a partial buffer against uncertain concentrate supplies. The Horne smelter's objective is to be recognized world-wide for its integrity, flexibility and technical excellence in sampling and processing a broad range of copper and PM bearing materials. An extensive program of testwork to continually improve expertise and treatment efficiency is thus underway. It may be expected that in an environmentally responsable society, this ability to recycle our primary resources will be of increasing value.

REFERENCES

1. G.C. McKerrow, "A 14 ft. x 32 ft. Converter at the Noranda Smelter", Pyrometallurgical Process in Non-Ferrous Metallurgy, edited by J.N. Anderson and P.E. Queneau, (The Metallurgical Society of AIME Conferences, Volume 39, Gorden and Breach Science Publisher, New York, N.Y., 1967, pp 247-258).

2. L.A. Mills, G.D. Hallett and C.J. Newman, "Design and Operation of the Noranda Continuous Smelting Process", Extractive Metallurgy of Copper, edited by J.C. Yannopoulos and J.C. Agarwal, (The Metallurgical Society of AIME, New York, 1976, pp. 458 - 487).

3. J.B.W. Bailey, G.D. Hallett and L.A. Mills, "The Noranda Smelter - 1965 - 1983", Advances in Sulfide Smelting, edited by H.Y. Sohn, D.B. George and A.D. Zunkel, (TMS-AIME, New York, N.Y., 1983, pp. 691-707).

4. P.J. Mackey, J.B.W. Bailey and G.D. Hallett, "The Noranda Process - An Update", Copper Smelting - An Update, edited by D.B. George and J.C. Taylor, (AIME, New York, 1981, pp. 213-236).

5. P. Tarassoff, Process R. & D. - The Noranda Process, (Metallurgical Transactions B, Vol. 15B, Sept. 1984, pp. 411-432).

6. J.B.W. Bailey and W.A. Dutton, Computer Control of the Noranda Continuous Smelting Process, (Paper presented at the 107th AIME annual meeting, Denver, Colorado, February 26th - March 2nd, 1978).

7. D.G. Pannell and P.J. Mackey, Noranda Process Operations 1988 and Future Trends, (Paper presented at the Copper Committee Meeting of the GDMB, Antwerk, Belgium, April 27-29, 1988).

8. D.A. Rawnsley, A. Crépeau and F. Marcil, "Shredded Scrap at the Horne Smelter", Proceedings of International Symposium on Recycled and Secondary Metals, (AIME, Fort Lauderdale, Florida, December 1985, pp. 563 - 573).

9. E. Palumbo et al., Sampling and Sample Preparation of PM Scrap at the Horne Smelter, (13th International Precious Metal Conference, IPMI, Montreal, Canada, June 11 - 15, 1989).

10. E. Palumbo, M. Bédard and C.H. Teh, Induction Furnace Sampling and Sample Preparation Methods, (Precious Metals Sampling and Analysis Seminar, IPMI, Oak Brook, Illinois, October 3-5, 1988).

11. C.H. Teh, J.B.W. Bailey and E. Palumbo, New Sampling Method for PM Scrap - Noranda Sampler, (13th International Precious Metals Conference, IPMI, Montreal, Canada, June 11-15, 1989).

12. M. Bédard, "Determination of Precious Metals in Scrap Samples by Fire Assay ICP Spectroscopy", Proceedings of the Tenth International Precious Metal Institute Conference, (Lake Tahoe, Nevada, pp. 119 - 136, 1986).

13. P. Giasson, D. Pinard and M. Bédard, Recent Improvements in Precious Metal Analysis, (13th International Precious Metal Conference, IPMI, Montreal, Canada, June 11 - 15, 1989).

14. A. Pelletier, J.M. Lucas and P.J. Mackey, The Noranda Tuyère Pyrometer - A New Approach to Furnace Temperature Measurement, (Paper presented at Copper '87, Vina del Mar, Chile, November 30th - December 3rd, 1987, ICM and CIM 1987).

15. F.W. Karasek et al, Analysis of Dust and Stack Samples from Noranda for Dioxins, Furans, PCB's and Other Organic Compounds, (University of Waterloo, Ontario, Canada, April 1986).

Simulation models for mineral processing plants

J.W. Merks
Matrix Consultants Limited, Vancouver, British Columbia, Canada

Abstract

Simulation is the process of designing a computerized model, of a system or procedure, for the purpose of understanding its behavior, and developing strategies to control its operation. Simulation models are rapidly becoming an effective tool for mineral processing plants. Powerful software and fast hardware are more readily available than a basic understanding of the theory of mathematical probability and applied statistics that process simulation models demand.

Simple spreadsheet software can be used to develop simulation models that take into account any set of conditions at a mineral processing plant. In addition to the set of variables on which two- and three-product formulas are based, a reliable variance estimate for each variable is required. How to estimate variances, effectively and at the lowest possible cost, and how to compute the composite variances for compound measurements such as metal contents, are key elements for useful simulation models.

Introduction

Simulation models for systems and procedures have become one of the most powerful and effective applications for computers (2,6). Simulations can be applied to a wide range of scientific and engineering disciplines. Generally, a model is designed to investigate the effects of changes in and interactions between a set of variables before a system is in operation or a procedure is implemented, by simulating weeks, months, and even years, in a few minutes of computer time.

Along with practical and useful applications always comes abuse and misuse. For a simulation process cannot be more reliable than the methodology that was applied to design the model, to check its performance, and to verify its validity. A mineral processing plant is an example of a dynamic system in which the variables changes continuously as a function of time but within probabilistic constraints that can be calculated from the variances for all variables in the set that determines the system.

The model can be designed to simulate not only the variables but also the variance for each variable. Models in which the variances for the set of variables are constants in the function that generates normally distributed random numbers, are less realistic than models in which the variables and variances are both simulated.

A deterministic model may apply to one subset of variables while a stochastic model may be more suitable to describe another. A system is deterministic if its present state is completely determined by a previous state, and stochastic if its present state cannot be predicted from a previous state with a high degree of certainty. Variables that interact in a mineral processing plant often exhibit deterministic and stochastic behaviors so that correlations between past and present states may reflect various degrees of significance. Dependencies between variables should be identified and taken into account.

Simulation models are based on mathematical probability and applied statistics (1,3,5). An important postulate from applied statistics is the Central Limit Theorem. This theorem dictates that the degree of precision with which the behavior of a dynamic system can be measured, increases with the number of measurements that are performed to interrogate it. The cost differential between a small set of real measurements and a very large set of simulated measurements essentially accounts for the popularity of simulation models.

Any computer with spreadsheet software is a suitable tool to simulate the operation of a mineral processing plant. The advantage of a spreadsheet is that the simulation model can easily be modified to reflect changes in process parameters, or better understanding of its operation. The simulation model can only be meaningful if it takes into account all interactions between the complete set of variables, and all possible effects of dependencies between any subset of variables.

The variance for the variable that contributes most to the system's intrinsic uncertainty, is the limiting factor for the precision of a simulation model which implies that the model can also be used to identify the most dominant variable in the process. How to obtain reliable variance estimates at the lowest possible cost, how to optimize simulation models most effectively, and how to compute a composite variance for a compound measurement such as a metal content, or a percentage recovery, are the key to powerful models for mineral processing plants.

Variances for Measurements

Variances for single measurements, and composite variances for compound measurements, play an essential role in effective simulation models for mineral processing plants. Under certain conditions the variances for different stages in a measurement chain are additive (3,5). For example, the sum of the sampling variance, the variance of sample preparation and the variance of analysis is often required in process simulations. However, the difference between two variances only becomes a meaningful estimate for the variance of an intermediate stage in the measurement chain if their F-ratio is statistically significant, and preferably significant to the extreme.

The Central Limit Theorem dictates that the variance for the mean of a set of elements (primary or secondary increments, subsamples or test samples, or assays) is equal to the variance between single elements divided by the number of elements in the set. This theorem underlies the general sampling formula, a simplified version of which follows:

$$var(t) = var(s)/n + var(p) + var(a)/k$$

in which:
- var(t) = total variance for a sampling unit
- var(s) = sampling variance
- var(p) = variance of preparation
- var(a) = variance of analysis
- n = number of increments/sampling unit
- k = number of measurements/test sample

Simulation models for mineral processing plants should take into account the effect of a serial correlation on the sampling variance (1,4). Measurements with an on-stream analyzer can be used to compute terms of the time series variances for each mass flow that is interrogated. The term that is closest to the time interval between primary increments is a reliable estimate of the sampling variance for the metal grade in this mass flow.

A significant serial correlation exists if the F-ratio between the variance for the completely randomized set of measurements and the first term of the time series variances for the ordered set exceeds the tabulated value at a certain probability level, and with the appropriate degrees of freedom. In the case that the calculated F-ratio is below the tabulated value the measurements in the set are randomly distributed, and thus statistically independent. Often, the sampling variance is significantly lower than the sum of the variances of sample preparation and analysis so that the serial correlation impacts only marginally rather than significantly on the total variance.

Sets of on-stream measurements for mass flows in mineral processing plants invariably display significant serial correlations. Time series variances for on-stream measurements are computed with the following formula:

$$var_i = \frac{\Sigma[x_{(j+i)} - x_j]^2}{2n - 1}$$

in which: var_i = variance term at ith spacing
x_{j+i} = measurement in (j+i)th increment
x_j = measurement in jth increment
i = spacing between increments
n = number of measurements in the set

A sampling variogram is a graph in which a set of time series variances is plotted against their spacings. If the term that matches the time interval between on-stream measurements is significantly lower than the variance for the completely randomized set, it becomes the most reliable estimate for the sampling variance. Dividing by the number of on-stream measurements in a set, and adding the variances of preparation and analysis for samples that are collected from the mass flows to the on-stream analyzer, generates an estimate for the total variance of a metal grade.

Generally, the total variance for the moisture content or metal grade of a sampling unit is computed by dividing the sampling variance by the number of primary increments in the set that constitutes a gross sample, and then adding the variance of sample preparation and the variance of analysis. The variance of preparation and analysis can be estimated from duplicate test samples, and the variance of analysis from assays of duplicate test portions, with the following formula:

$$var(x) = (\pi/4) * [\Sigma(|x_{1j} - x_{2j}|)/k]^2$$

in which: $var(x)$ = variance for a single measurement
x_{1j} = first measurement on jth pair
x_{2j} = second measurement on jth pair
k = number of measurements in the set

A cost effective method to estimate the total variance for the metal grade of a mass flow is to collect a pair of interpenetrating gross samples from a set of no less than four (4) sampling units. The advantage of such a sampling regime is that it takes into account, almost quantitatively, the effect of a serial correlation on the sampling variance. After all, the sampling variance for each interpenetrating gross sample is, in effect, equal to the second term of the time series variances rather than the first term.

Ideally, the variance of analysis should be the dominant component in the process of sampling, preparation and analysis. In the case of crushed ore the sampling variance and the variance of sample preparation are usually higher than the variance of analysis. In the case of ball mill discharge, cyclone overflow, tailings and concentrates the sampling variance is only a fraction of the variance of analysis. Moreover, the variance of preparation can easily be reduced to match the variance of analysis. Rod mill discharge is expensive to sample mechanically, and difficult to sample manually in an unbiased manner and with an acceptable degree of precision.

Concentrates in thickeners contribute a large measure of uncertainty to production and inventory, and thus large variance components to simulation models for mineral processing plants. Measuring the wet mass of concentrate with a static scale after dewatering or drying reduces the variance for the wet mass to the lowest possible level. Mineral concentrates can be sampled for moisture contents and metal grades with a high degree of precision. In fact, sampling and weighing of concentrate after dewatering or drying at the mineral processing plant generates a point of reference against which all production and smelter data should be reconciled.

The variance for the wet mass of crushed ore that is measured with a belt scale, can usually be estimated at low cost. During a maintenance period the conveyor with the belt scale could be operated, either with its chain in the proper position, or with a static load applied to its frame, for intervals of 30 - 60 min. After each interval the measured mass is recorded, and this process is repeated until no less than four (4) measurements are obtained. The variance for this set can then be used to estimate the variance for the wet mass that is processed during a production period.

The set of calibration data for a static scale provides sufficient information to estimate the variances at gross and tare loads, and thus the variance for a net wet mass in a rail car, truck or bulk bag. The sum of the variances for any number of units becomes the variance for the cumulative wet mass which can then be used in the simulation model.

Belt scales are perfectly acceptable to measure the wet mass of crushed ore to a mineral processing plant, or to a leach pad, but not precise enough to measure the wet mass of a concentrate during its transfer to an inventory at a mine. In particular if a large inventory is retained for a long period of time, all random variations and systematic errors in the mass measurement process accumulate, and often cause a high degree of uncertainty in such an inventory. Simulation models can also be applied to assess the effect of random variations on the precision for concentrate inventories.

Composite Variances for Compound Measurements

A simulation model is based on a set of variables that each displays a degree of variability for which the variance is the basic measure. All variables that interact in a mineral processing plant are continuous but constrained within probabilistic limits. Reliable, or at least realistic, variance estimates for wet mass, moisture content and metal grades of mill feed, and for metal grades of concentrate and tailings are required to compute composite variances for dry mass, metal content and percentage recovery.

The variances for a set of stochastic variables in a function interact in a deterministic model that finds its origin in calculus, and more particularly in partial derivatives of multivariate functions. The composite variance for any multivariate function is equal to the sum of the squared partial derivative for each variable in the set multiplied by its variance. In formula:

$$\sigma^2(y) = \left(\frac{\partial y}{\partial x_1}\right)^2 \sigma^2(x_1) + \left(\frac{\partial y}{\partial x_2}\right)^2 \sigma^2(x_2) + \ldots + \left(\frac{\partial y}{\partial x_n}\right)^2 \sigma^2(x_n)$$

The above equation is based on the general formula so that it applies to all functions. The sigma symbols imply unknown population variances which are replaced by estimates from samples in practical applications. The set of variables should be independent. Otherwise, the effect of a correlation between any subset of variables should be taken into account.

For complex systems such as a mineral processing plant with a heavy medium plant and flotation circuits for lead and zinc, a deterministic model is virtually impossible to develop. In fact, complex systems are the very reason why simulation models have become such a powerful tool. A simple model will be used to show how to compute a composite variance, how to use it in a simulation model, and how to check its validity.

The metal content in metric tons of a quantity of crushed ore, concentrate, or tailings, is calculated from the wet mass in metric tons, and the metal grade and moisture content in percent, with the following formula :

$$Me = wM * AF * MF$$

in which: Me = metal content in mt
wM = wet mass in mt
AF = metal factor: %A/100
%A = metal grade in percent
MF = moisture factor: $(100 - \%H_2O)/100$

Based on the formula for the composite variance of a multivariate function the partial derivatives for this elementary function are:

$$\left(\frac{\partial Me}{\partial wM}\right) = AF * MF \qquad \left(\frac{\partial Me}{\partial AF}\right) = wM * MF \qquad \left(\frac{\partial Me}{\partial MF}\right) = wM * AF$$

The sum of the products of squared partial derivatives and variance estimates is equal to the composite variance for the metal content in metric tons so that the following formula applies:

$$var(Me) = (AF * MF)^2 \, var(wM) + (wM * MF)^2 \, var(AF) + (wM * AF)^2 \, var(MF)$$

in which the substitution of σ^2 with var(x) implies that unknown population variances are replaced with variance estimates that are measured in samples.

Following is a numerical example to show how the composite variance for a metal content is computed. A wet mass of 10,000 mt mill feed was measured with a belt scale for which a coefficient of variation (CV in %) of 1.6% was reported. The variance of sampling, preparation and analysis for a copper grade of 0.53%, and for a moisture content of 3.2%, were measured with CV's of 2.5% and 5.0% respectively.

The coefficient of variation is equivalent to the standard deviation as a percentage of the measured variable so that the variance for a wet mass of 10,000 mt is: $(10{,}000 * 1.6/100)^2 = 25{,}600\ mt^2$, the variance of sampling, preparation and analysis for a copper grade of 0.53% is: $(0.53 * 2.5/100)^2 = 0.000{,}176\ \%^2$, and the variance of sampling, preparation and analysis for a moisture content of 3.2% is: $(3.2 * 5.0/100)^2 = 0.0256\ \%^2$. Based on the variances for wet mass, metal grade and moisture content the components of the composite variance for a copper content of: 10,000 * (0.53/100) * (100 - 3.2)/100 = 51.304 mt are:

WET MASS	: $(0.0053 * 0.968)^2 * 25{,}600$	=	0.6738
COPPER GRADE	: $(10{,}000 * 0.968)^2 * 0.000{,}176 * 10^{-4}$	=	1.6492
MOISTURE CONTENT	: $(10{,}000 * 0.0053)^2 * 0.0256 * 10^{-4}$	=	0.0072
COPPER CONTENT	:		2.3302

Multiplication with 10^{-4} accounts for the fact that metal and moisture factors rather than percentages are used to compute their variance contributions to the composite variance of 2.3302 mt^2. This variance of: var(Cu) = 2.3302 mt^2 translates into a standard deviation of: sd(Cu) = $\sqrt{2.3302}$ = 1.5265 mt, a 95% confidence interval (95% CI) of: z0.96 * sd(Cu) ≈ 2.0 * 1.5265 = ± 3.053 mt, or: 3.053 * 100/ 51.304 = ± 6.0%, and a 95% confidence range (95% CR) from: 51.304 - 3.053 = 48.3 mt up to: 51.304 + 3.053 = 54.4 mt. The factor z0.95 ≈ 2.0 is the rounded value of 1.96 for the symmetrical 95% probability from the Gaussian or normal distribution.

The components of the composite variance for a copper content of 51.3 mt show that the measurement of wet mass contributes: 0.6738 * 100/2.3302 = 29%, that the measurement of grade adds: 1.6492 * 100/2.3302 = 71%, and that the measurement of moisture accounts for: 0.0072 * 100/2.3302 = 0.3% only. Hence, the composite variance can be reduced, and thus the precision for copper content improved, by reducing the total variance for metal grade. If the sampling variance, and the variances of preparation and analysis, were known, then the total variance for the measurement of copper in mill feed can be optimized most effectively.

In this case the wet mass, metal grade and moisture content are expected to vary independently which implies that correlations between these variables will not occur. However, if the stickiness of crushed ore were a function of its moisture content, then the precision of the belt scale too could become a function of moisture content. Under such a condition stickiness would not only result in a higher variance for the wet mass of mill feed but could also cause a bias or systematic error.

The precious metal content in kilograms of a quantity of ore or concentrate is calculated from the wet mass in metric tons, the precious metal grade in kg/mt, and the moisture content in percent. In formula:

$$Me = wM * A * MF$$

in which: Me = precious metal content in kg
wM = wet mass in mt
A = metal grade in kg/mt
MF = moisture factor: $(100 - \%H_2O)/100$

The wet mass of 25,000 mt crushed ore was measured with a belt scale for which a CV of 2.1% was reported. The variance of sampling, preparation and analysis for a gold grade of 3.2 g/mt, and a moisture content of 2.4%, were measured with CV's of 3.6% and 4.5% respectively.

The variance for the wet mass is: $(25,000 * 2.1/100)^2 = 275,625$ mt^2, the variance for the gold grade is: $(3.2 * 3.6/100)^2 = 0.013,271$ g^2/mt^2, and the variance for the moisture content is: $(2.4 * 4.5/100)^2 = 0.011,664$ $\%^2$ so that the components of the composite variance for a gold content of: 25,000 * 3.2/1,000) * (100 - 2.4)/100 = 78.08 kg are:

WET MASS	:	$(0.0032 * 0.976)^2$	$* 275,625$	= 2.6886
GOLD GRADE	:	$(25,000 * 0.976)^2$	$* 0.013,271 * 10^{-6}$	= 7.9010
MOISTURE CONTENT	:	$(25,000 * 0.0032)^2$	$* 0.011,664 * 10^{-4}$	= 0.0075
GOLD CONTENT	:			10.5970

Multiplication with 10^{-6} accounts for the fact that a gold grade of 0.0032 kg/mt rather than 3.2 g/mt was used to calculate the grade component of the composite variance for gold content. Thus the variance: var(Au) = 10.5970 kg^2 is equivalent to: sd(Au) = $\sqrt{10.5970}$ = 3.2553 kg, a 95% CI of: $z0.96$ * sd(Au) ≈ 2.0 * 3.2553 = ± 6.5106 kg, or: 6.5106 * 100/ 78.08 = ± 8.3%, and a 95% CR from: 78.08 - 6.51 = 71.6 kg up to: 78.08 + 6.51 = 84.6 kg.

These components of the composite variance show that the measurement of wet mass contributes: 2.6886 * 100/10.5970 = 25%, that the measurement of metal grade adds: 7.9010 * 100/10.5970 = 75%, and that the measurement of moisture content accounts for: 0.0075 * 100/10.5970 = 0.1% only. Hence, the composite variance can be reduced, and thus the precision for gold content improved, by reducing the total variance for the gold grade.

If the sampling variance and the variances of preparation and analysis were known, the total variance for the measurement of gold in crushed ore can be optimized most effectively. Due to the single particle or nugget effect the variance of analysis is often the largest component of the total variance. The variance of analysis can be reduced by screening test samples at 100 - 150 mesh, and assaying coarse and fine fractions separately.

Mechanical sampling systems should preferably be designed to collect pairs of interpenetrating gross samples. Without a mechanical sampling system pairs of interpenetrating gross samples should be collected from time to time to obtain a reliable estimate for the total variance of the metal grade, and to ensure that the manual sampling procedure is optimized. For an effective simulation model requires reliable variance estimates for the most dominant variables, and realistic variance estimates for all others.

Normally Distributed Random Numbers

Spreadsheets provide pseudo-random number generators that are based on a standard uniform distribution within an interval of: $0 \leq r_i \leq 1$. The term "pseudo-random" implies that the same seed will always generate the same sequence of random numbers. More advanced random number generators are reseeded with an internal variable from the microprocessor each time the generator routine is called to ensure that the probability for periodicities to occur is greatly reduced.

The Central Limit Theorem implies that the sum of n identically distributed independent random variables approximates a normal distribution with a mean of $n\mu$ and a variance of $n\sigma^2$ where μ and σ^2 are the mean and the variance of the variable. If the variables X_i, i = 1, 2, ... , n, follow the standard uniform distribution, then $\mu = 0.5$, and $\sigma^2 = 1/12$. Hence, the sum of a set of n random numbers from the standard uniform distribution approximates a normal distribution with a mean of $0.5n$ and a variance of $n/12$.

The choice of n in the process of generating normally distributed random numbers is essentially a matter of computing efficiency. The larger the number of uniformly distributed random numbers in a set, the more closely the mean approaches the normal distribution. Nevertheless, a set of twelve standard uniform random numbers provides a realistic approximation of the normal distribution, and eliminates the need to divide the variance by the factor 12 in order to transform the distribution from nonstandard normal to standard normal. Based on these considerations the following formula is the most efficient choice to simulate mineral processing plants:

$$x(r) = x(o) + var(x)^{\frac{1}{2}} * \left(\sum^{j=12} r_j - 6\right)$$

in which: $x(r)$ = simulated measurement
$x(o)$ = observed measurement
$var(x)$ = variance estimate for $x(o)$
r_j = jth standard uniform random number

A simple example will be used to demonstrate how to generate normally distributed random numbers. The sum of twelve standard uniform random numbers that were generated in sequence with the function "@RAND" in a spreadsheet, turned out to be 4.9123. If a wet mass of 25,000 mt were measured with a variance of 275,625 mt^2, a normally distributed random number for this wet mass would be: 25,000 + $\sqrt{275{,}625}$ * (4.9123 - 6) = 25,000 + 525 * (- 1.0877) = 24,429 mt. Another sum of twelve standard uniform random numbers was 6.1462, and resulted in a normally distributed random number of: 25,000 + 525 * 0.1462 = 25,077 mt.

Based on a variance of 275,625 mt^2, and a z-value of 1.96, the 95% CI for the wet mass of 25,000 mt is: 2.0 * $\sqrt{275{,}625}$ = ± 1,050 mt for a 95% CR range from: 25,000 - 1,050 = 23,950 mt up to: 25,000 + 1.050 = 26,050 mt. Since randomly generated wet masses of 24,429 mt and 25,077 mt fall within the 95% CR for the observed wet mass of 25,000 mt, each appears to be an unbiased estimate for the unknown true mass of the quantity of crushed ore.

The wet mass of 25,000 mt and its variance of 275,625 mt^2 can also be used to generate variance estimates by simulation. After all, the variance too is a stochastic variable that is precise within predictable probability limits. Any set of normally distributed random numbers can be used to compute variance estimates but sets of twelve will be used in numerical examples.

The validity of simulation models for variances can be verified by applying statistical tests. For example, a set of twelve variance estimates, each based on a set of twelve normally distributed random numbers, resulted in an average variance of 229,111 mt^2. The F-ratio of: 275,625/229,111 = 1.20 between the observed variance of 275,625 mt^2 and the simulated average variance of 229,111 mt^2 is below the tabulated values of $F0.95;\infty;132$ = 1.24 and $F0.99;\infty;132$ = 1.36 which implies that these variances are statistically identical, and thus compatible. Hence, the pseudo-random number generator is unbiased for small sequences.

Any set of variances that is obtained with this standard uniform random number generator should be homogeneous, and each set should be compatible with the variance that is based on a deterministic model. Statistical tests that check sets of variances for homogeneity such as Bartlett's chi-squared test, and pairs of variances for compatibility such as Fisher's F-test, should be applied to validate simulations that are based on the "@RAND" function in spreadsheet software.

Homogeneity of Variances

The wet mass, metal grades and moisture content of a quantity of ore, and the variances for this set of variables, were used to compute the composite variances for copper and gold contents. In Table I are listed all variables and variances that were used to compute the composite variances:

TABLE I
Variables and Variances for Copper and Gold Contents

Parameter	Symbol	Cu in mt	Au in kg
Wet Mass in mt	wM	10,000	25,000
Variance in mt^2	var(wM)	25,600	275,625
Moisture Factor	MF	0.968	0.976
Variance	var(MF)	$0.0256 * 10^{-4}$	$0.011,664 * 10^{-4}$
Copper Factor	AF	0.0053	
Variance	var(AF)	$0.0176 * 10^{-6}$	
Gold Grade in kg/mt	A		0.0032
Variance in kg^2/mt^2	var(A)		$0.013,271 * 10^{-6}$
Copper Content in mt	Cu	51.304	
Composite Variance in mt^2	var(Cu)	2.3302	
Gold Content in kg	Au		78.08
Composite Variance in kg^2	var(Au)		10.5970

The sets of variables and variances in Table I, and the formulas for the composite variances, were used to simulate the composite variances of 2.3302 mt^2 for a copper content of 51.3 mt, and 10.5970 kg^2 for a gold content of 78.1 kg. The first step is to generate variance estimates for the wet mass in mt, the copper factor, the gold grade in kg/mt, and the moisture factor, with the formula: $var[x(r)] = var[x(o) + (\Sigma r_i - 6) * \sqrt{var(x)}]$ so that each variance estimate is based on twelve normally distributed random numbers.

The second step is to generate a single normally distributed random number for each variable from the simulated variance and the observed measurement so that: $x(r) = x(o) + (\Sigma r_i - 6) * \sqrt{var[x(o)]}$. Finally, sets of simulated variables and variances are used to compute an estimate for the composite variance of the metal content. The process is repeated until a set of twelve estimates for the composite variance are obtained. In Table II are listed twelve estimates for the composite variances for copper and gold contents:

TABLE II
Simulated Composite Variances for Metal Contents

No	var(Cu)	var(Au)	No	var(Cu)	var(Au)
1	2.0269	18.1243	7	2.1693	10.1094
2	1.8163	10.9280	8	2.3151	11.7346
3	3.5187	12.4773	9	1.3230	6.2503
4	1.3867	7.0809	10	1.1189	12.9080
5	3.5203	9.7642	11	3.5414	7.2351
6	2.5627	9.7417	12	1.6856	12.3506
Average Composite Variance				2.2487	10.7254
Computed Composite Variance				2.3302	10.5970

Fisher's F-ratio test is applied to verify that F-ratios between computed and simulated variances are below tabulated F-values. The calculated ratios of:

2.3302/2.2487 = 1.04, and: 10.7254/10.5970 = 1.01, are so close to unity that computed and simulated variances are statistically identical. The question remains whether the sets of composite variances are homogeneous.

Bartlett's chi-squared test is applied to check whether sets of simulated composite variances are homogeneous, and thus whether the simulation model is valid. The chi-squared test is based on the following formula:

$$\chi^2 = \Sigma(k_j - 1) * \ln[var(\bar{x})] - \Sigma[(k_j - 1) * \ln[var(x_j)]]$$

in which:
- χ^2 = calculated chi-squared value
- $\ln[var(x_j)]$ = natural logarithm for jth variance
- $\ln[var(\bar{x})]$ = natural logarithm for average variance
- k_j = number of measurements in jth set

A comparison of the calculated chi-squared values of 9.264 for copper and 5.359 for gold with the tabulated values from the chi-squared distribution at 11 degrees of freedom reveals that the probability for 9.264 falls between 8.148 at 70% and 10.341 at 50%, and that the probability for 5.359 falls between 4.575 at 95% and 5.578 at 90%. Since those confidence ranges for calculated chi-squared values are neither too low nor too high, there is no evidence that the simulation model is invalid.

Possible effects of periodicities in the "@RAND" function can be assessed by increasing the number of normally distributed random numbers in each set to match the number required for a particular simulation model. In the case that a set of variances for large sequences of normally distributed random numbers is homogeneous, the standard uniform distribution of pseudo-random numbers that the "@RAND" function provides is also reliable for simulation models that require large sequences. However, if the periodicities in the "@RAND" function and the required sequences were to coincide, the random number generator should not be applied to process simulations.

Summary

Spreadsheet software can be used to develop advanced simulation models for mineral processing plants. The pseudo-random number generator for the standard uniform distribution that spreadsheet software routinely provides should be tested to check its performance and suitability for simulation models. Applied statistics provides tests to check performance by testing pairs of variances for compatibility, and sets of variances for homogeneity.

Models that simulate all variables and variances that interact in a mineral processing plant are more realistic and effective than models that simulate variables from constant variances. Reliable variance estimates are the key to meaningful simulation models for mineral processing plants. Unless variables and variances are generated simultaneously the simulation model for a mineral processing plant cannot truly reflect its behavior. Nor does it generate a reliable variance estimate for the mass of tailings, and, by implication, for the mass of concentrate that is transferred to a thickener or an inventory.

Sets of on-stream measurements can be used to compute terms of time series variances, and to estimate the sampling variance for each metal grade. Pairs of interpenetrating samples are useful to estimate the variance of sampling, preparation and analysis for moisture contents and metal grades in crushed ore and concentrate, and for metal grades in wide range of mass flows.

The effect of dependencies between subsets of variables can be taken into account by conditional simulation on the basis of correlation-regression parameters between dependent and independent variables. The metal grades in mill feed and tailings are frequently correlated so that the simulation model should reflect that the metal grade in tailings is a function of the metal grade in mill feed.

The arithmetic mean of a set of on-stream measurements for the metal grade of a mineral concentrate is often biased due to the fact that low mass flows in flotation cells cause with higher metal grades in concentrate, and high mass flow cause lower metal grades in concentrates. The weighted average that is calculated from the sum of products of all metal grades and mass flows for a production period, is expected to be unbiased.

Simulation models can be applied to a simple mineral processing plant that can be described with the two-product formula, and to a complex operation with a heavy medium plant, an intermediate concentrate, and lead and zinc concentrates. Simulation models provide a more profound understanding of interactions between variables and variances in a mineral processing plant than daily or monthly metallurgical balances.

References

1. Box, G E P and Jenkins, G M, Time Series Analysis
Holden-Day, San Francisco, California, 1976

2. Graybeal, W and Pooch, U W, Simulation: Principles and Methods
Winthrop Publishers, Inc, Cambridge, Massachusetts, 1980

3. Merks, J W, Sampling and Weighing of Bulk Solids Transtech Publications, Clausthal-Zellerfeld, 1985

4. Merks, J W, The Effect of Serial Correlations on the Sampling Variance
CIM Conference, Canadian Mineral Processors Division
Computer Applications and Process Control, Paper 44, 1985

5. Volk, W, Applied Statistics for Engineers
R E Krieger Publishing Company, Huntington, New York, 1980

6. Yakowitz, S J, Computational Probability and Simulation
Addison-Wesley Publishing Company, Inc, Reading, Massachusetts, 1977

Mechanical feed systems — in-line weighing/accounting

C. Karpowicz and T. Fahlenbock
Control and Metering Limited, Toronto, Ontario, Canada

1.0 Introduction

To compete profitably against third world country production, today's metallurgy industry requires an even tighter control over the processing of raw materials. To meet this end, Canadian mineral processors must achieve cost effective utilization of raw materials to reduce costs and produce higher grade products to command a higher selling price. Today, this is achievable due to the advancement of the overall process control system (in conjunction with the accurate metering of raw materials) to quickly correct off-spec products on line to produce a consistently high grade.

This paper discusses several types of volumetric dispensing (metering) feeders and gravimetric (by weight) control techniques explaining not only the principles of operation but also the selection criteria to enable a user to choose the best combination for the application. Screw, vibrating tray and belt feeders are presented with both loss in weight and weigh belt control techniques. Several examples of actual installations at Canadian mineral processing plants will be highlighted.

2.0 Installation Profiles

In preparation for this paper, we contacted a few users of feeding equipment to learn the application and the results. The information below summarizes some of these situations.

A zinc processor feeds aluminum chips to a zinc alloying process prior to a coating line. Loss in weight feeders using vibrating tray feed mechanisms are used and have replaced weigh belt feeders and achieved feed rate accuracies of better than $\pm 1\%$ of actual rate. Feed rates vary between 6 kg/hour and 180 kg/hour. The vibrating tray feed mechanism is virtually maintenance free and the loss in weight feeder is self calibrating.

An aluminum processor uses a loss in weight feeder to feed Beauspar at rates up to 2.7 metric tonnes per hour. Beauspar is reclaimed waste product in the form of a fine powder weighing 43 lbs./cu.ft. to 49 lbs./cu.ft. Due to the poor flowing properties of the powder, the 15 cu.ft. weigh hopper is agitated to ensure an even flow into the 4 inch diameter metering screw.

Another aluminum company uses 14 inch belt feeders to meter aluminum fluoride at rates up to 2 metric tonnes per hour. Aluminum fluoride is a free flowing powder, 60 to 70 lbs./cu.ft. and feed rate accuracies better than $\pm$1% of actual rate compared to set point are achieved. The belt feeders are retared once per month.

A nickel processor feeds zinc powder into the zinc purification stage of the process to purify the final zinc solution being produced. Feed screws are used to meter the feed rate. Excellent feed rate accuracy is achieved, however, maintenance is fairly high due to the abrasive and smearing nature of the zinc powder. Seals wear prematurely necessitating this increased maintenance. The end user and the manufacturer are jointly experimenting together and a solution appears close at hand.

3.0 Volumetric Screw Feeders and Vibrating Tray Feeders

The above are but a few examples of feeder installation where weight control is used to precisely meter ingredients into the process.

Gravimetric feeders were selected in these cases in order to enter a feed rate in weight units and to achieve the precise control.

Volumetric feeders are often used to meter dry ingredients. Two popular types are screw feeders and vibrating tray feeders. Both dispense a certain volume of ingredient per unit of time. The screw feeder can have its feed rate varied by varying screw speed. The relationship between screw rpm and delivered volume is linear up to a certain maximum speed, above which the screw flights will not fill completely and feed rate linearity drops off. At very low rpms, the dispensed ingredient does not deliver in a uniform stream, it tends to dispense non uniformly. Normal single screw speeds vary from 10 rpm to 200 rpm for screws up to about 6 inches diameter.

Vibrating tray feeders normally function at constant frequency and variable amplitude. The feed rate variation is not linear with amplitude variation control since tray loading can affect the amplitude. A recent control development incorporating an amplitude sensor and feed back control produces a linear response over a very wide feed range (50 to 1).

4.0 Gravimetric Feeders

A gravimetric feeder is a device which can accept a setpoint by weight and feed material to the process at a controlled rate around that setpoint. This is in contrast to a volumetric feeder which cannot accept a setpoint by weight and the output must be estimated based on a historical calibration curve which may or may not be accurate.

4.1 Loss In Weight Feeders

A LIW feeder has four basic components as follows, ignoring for the time being the refill device (refer to Figure 1 - Components of a Continuous Loss in Weight Feeder):

a) Scale - a scale is a device which can measure and display the weight of a load while containing and supporting it. The precision of the scale is a function of the accuracy of weight measurement required and the stabilization time available between load change and measurement reading. Depending on the particular manufacturer's design, scales can be suspended or platform type, mechanically or electronically tared, analog or digital, and high or low displacement.

b) Feed Mechanism - is mechanically identical to a volumetric feeder as described in 3.0, but when used in this context it is mounted on the scale and used as a "take away" device to dispense ingredient at a controlled, accurate rate into the process. The feed mechanism must be equipped with variable speed drive. The proper feed mechanism is usually determined from experience (plant and supplier), accuracy requirements and testing.

c) Weigh Hopper - stores a pre-determined amount of ingredient (typically, 4-5 minutes worth) directly above the feed mechanism for introduction to the process. As a result it is mounted as an integral package with the feeder. Care must be taken to supply the hopper with the proper geometry and agitation (if required) to ensure the material does not bridge or rathole and flows in a uniform manner into the feed mechanism.

d) Feeder Controller - utilizes a PID based algorithm to accept a setpoint input (whether local or remote), compares the actual feed rate to the setpoint, generates a control signal output to the feed mechanism to maintain or change the motor speed (or vibration amplitude) and stores and makes accessible the total weight of material fed over the previous time period. Modern day processes generally require the feeder controller(s) to tie in with the overall plant control system. Some ways this is done:

   i) local control direct from controller keypad, perhaps with hand wired interlocks;

ii) remote control, usually utilizing a PLC interface;

iii) central control from graphics screen including data manipulation and software SPC packages.

By combining the above four components we now get an idea how a LIW feeder operates. Picture a 10 cu.ft. hopper on a mechanically counterbalanced scale full of material weighing 100 lbs./cu.ft., resulting in a scale weight of 1,000 lbs. If, at 9:00 a.m., the feeder is started at a setpoint of 100 lbs./hr., then the scale weight should be 900 lbs. at 10:00 a.m. If the weight is, say, 902 lbs., then the scale has "lost" only 98 lbs. so the control signal would increase the speed of the feed mechanism to get back to setpoint. In reality, of course, weight signals are sampled not hourly, but every half second or so to permit continual feed rate adjustment and maintain process control.

After some time period, the hopper will run out of material and will require refilling. It is clear, however, that the LIW algorithm cannot operate while the system is gaining weight rather than losing it. The solution is to refill the weigh hopper as quickly as possible (compared to the gravimetric feeding time) while maintaining a constant speed of the feed mechanism calculated by averaging the last few control signal outputs just prior to refill. When the pre-selected high hopper level is reached the refill device shuts off and after a scale stabilization period the feeder reverts to gravimetric control.

The refill device must be isolated from the feeder via a flexible connection and must have a capacity of at least 15-20 times the maximum feed rate of the LIW feeder to minimize the time spent in refill mode. Refill should occur in less than 10 seconds, if possible. The refill device is generally supplied by the end user but can be furnished by the feeder manufacturer if required.

4.2 Weigh Belt Feeders:

A weigh belt feeder works on a different principle to that of a LIW to achieve the same result; it measures true weight rather than a difference in weight. A weigh belt feeder is located directly under a supply hopper or bin without the requirement of a refill device. When the feeder is started the material is sheared under an inlet gate (set to an optimum pre-determined height) and travels over a weighed section consisting of a cantilevered weigh bar or weigh deck. The principle is that the feed rate is proportional to the belt load multiplied by the belt speed where the proportionality constant is a function of the length of the weighed section. Or, to put it another way:

$$\text{weight per foot } \left(\frac{\text{lbs.}}{\text{ft.}}\right) \times \text{speed } \left(\frac{\text{ft.}}{\text{min.}}\right) = \text{feed rate } \left(\frac{\text{lbs.}}{\text{min.}}\right)$$

The feeder controller accepts the two measured variables above (weight signal from the load cell and speed signal from a tachometer), integrates them, calculates a feed rate, compares it to the setpoint and sends a corrective signal to the motor speed controller. In this case the belt load is constant (assuming relatively constant bulk density) and the belt speed is variable to maintain process accuracy.

A belt feeder can also be used as a meter to measure a delivered flow to the belt weigher. In this situation the belt load is variable and the belt speed is constant since we are not setting a feed rate but simply measuring a wild flow across the unit. The belt meter has a totalizer output for inventory purposes and also includes an analog output proportional to feed rate for slaving other feeders if required. Keep in mind that a belt feeder can always be a meter, but a meter may not necessarily be able to function as a feeder depending on the original vendor's scope of supply.

As with volumetric and LIW feeders, there are several guidelines for optimal operation of weigh belt feeders. The sprocket ratios and gate height should be selected such that the feeder operates in a linear range of the load cell and at approximately 75% of maximum motor speed at the maximum feed rate. This allows maximum flexibility in terms of material bulk density changes and turndown requirements.

Automatic belt tracking and tensioning are key design requirements when specifying weigh belt feeders. An inlet slide gate should be placed between the feeder and overhead supply bin to allow maintenance and calibration of the feeder. A sampling valve should be placed in the discharge stream to divert the process material into a sampling container and provide the most accurate calibration.

Finally, pre-feeders should be utilized directly upstream of weigh belt feeders for dusty or floodable materials that would flood uncontrollably onto a belt placed directly under a bin. The pre-feeder is simply a volumetric device as previously discussed (e.g., screw, rotary valve, vibrating tray in many cases) equipped with variable speed control and serves to feed the process material in an orderly fashion onto the belt.

5.0 Feeder Selection

Now that we have discussed the various types and designs of feeders available we can begin to examine the evaluation process for selecting the optimum unit for a given process requirement. This will be done in 3 steps, as follows:

Step 1 - We will select the appropriate feed mechanism (whether volumetric or gravimetric) as a function of material characteristics.

Step 2 - We will decide between volumetric and gravimetric (LIW or weigh belt) feeding.

Step 3 - Assuming the decision in Step 3 above was in favour of gravimetric, we will select between LIW and weigh belt feeders.

Step 1: Feed Mechanism vs. Material:

The following comments will serve to summarize common feed mechanisms:

a) Single Screws:

- available from all feeder manufacturers up to 6" diameter (approximately 900 cu.ft./hr. capacity) with some offering up to 12", 16", or even 29" screws;

- two main types, open spiral and blade (sometimes termed solid flight);

- output generally linear with speed, i.e., doubling the rpm doubles the feed rate;

- pulsing discharge pattern;

- moderately priced compared to other feed devices.

i) Spiral Screws:

- for pellets and free flowing powders and granules (i.e., salt);
- use over-sized tube when feeding pellets to prevent screws jamming in tube;
- large diameter spiral screws are good for materials which stick to screws, the smaller surface area of a spiral screw vs. a blade permits less material adherence to the screw;
- do not use for low feed rates of stocky powder - even a slight buildup drastically decreases volumetric capacity;
- do not use for floodable powders;
- do not use for heavy metal powders as the coil may bend, in certain applications spiral screws made of thick, heady duty stock can be used;
- consider releif grinding the screw when powder builds up on a hard layer on the inside of the screw tube, this reduces friction by reducing the screw surface area in contact with the tube.

ii) Blade Screws:

- use for most powders which do not stick to metal;
- can control materials with a slight tendency to flood if refill level in hopper is high and hopper venting is adequate;
- be careful with materials such as Titanium Dixoide - capacity drops with buildup;
- use when strong screw is required, i.e., for heavy powders or extended length screw with outboard bearing support.

b) Twin Screws:

- available from a few manufacturers only, up to the 80 mm diameter size range with a maximum output in the area of 180 cu.ft./hr. (for twin concave);
- four main types - twin concave (self cleaning), twin blade, twin spiral and double spiral;
- output linear with speed;
- discharge also pulsing, but less than a single screw;
- high priced compared to single screws and other feed devices; relative price difference increases with capacity.

i) Twin Concave Screws:

- ideal for floodable and cohesive powders; twin concave screws closely intermesh and rotate in the same direction, forming advancing pockets which fill with material and seal against leakage;
- good for use with powders that stock to metal; screws will self clean after initial layer is built up;
- do not use for metal powders - too much friction in close screw clearances;
- do not use for powders containing hard lumps, screws will jam - follow manufacturers recommendation for maximum allowable non-crushable particle size;
- big accuracy advantage over single screws at low feed rates - twin concaves run at higher speeds than comparatively sized single screws, and there are two discharge points rather than one - this produces a more even flow critical for low rate accuracy.

ii) Other Twin Screws:

- twin spirals and blades similar in application to equivalent single screws;

- very convenient for applications requiring a variety of materials to be fed with the same feeder - it at least one of the materials requires a twin concave, screws can easily be interchanged for flexibility in feeding other materials not suitable for a twin concave.

c) Vibrating Trays:

- offered by most manufacturers up to approximately 12" wide, with larger trays available upon request;
- output is not linear with vibration amplitude unless amplitude feedback is provided;
- steady, non-pulsing discharge;
- comparatively low priced;
- good for free flowing pellets and granules;
- very useful for fibers and flakes which may be damaged by a screw feeder;
- heavy duty trays good for use on abrasive material which may cause excessive screw wear;
- do not use for floodable materials or materials with a tendency to pack;
- range 50:1 with amplitude feedback;
- difficult to predict actual flow rate achievable with a particular tray on a particular material - guidelines are available, but testing should be done to be safe.

d) Rotary Valve (Airlocks):

- wide range of sizes available;
- output linear with speed assuming pockets fill and empty repeatably;
- moderately priced, depending on design requirements;
- excellent for pre-feeding floodable materials onto weigh belt feeders;
- limited use as LIW feed mechanisms due to pulsing nature of vanes, pockets not emptying completely, etc., however, good for high rate, floodable applications where accuracy is not critical or can be combined with a screw for more stringent accuracy requirements.

e) Belts:

- can be used as a stand-alone- weigh belt feeder, or a volumetric feed mechanism mounted on a LIW scale.
- available from most feeder companies, generally in 2 or 3 basic sizes ranging from a 6 inch wide belt to a 30 inch wide belt - wider units are available for large, bulk material handling applications;

Step 2 - Gravimetric vs. Volumetric

The determination of the proper feed mechanism as discussed in Step 1 is the first and arguably the most critical step in the feeder selection process. If the feed mechanism is not correct and/or it does not receive a steady flow of material from upstream hoppering devices then no feeder, no matter the degree of electronic sophistication, will work to the user's satisfaction.

Assuming we have now selected the proper feed mechanism, we now arrive at the issue of volumetric vs. gravimetric feeding. The following factors will be considered:

a) Accuracy requirements.

b) Information availability.

c) Fault detection.

d) Continuous mixture control.

e) Price.

a) Accuracy Requirements - to simplify the discussion, accuracy will be discussed in terms of repeatability only since this is usually what is implied in an incomplete accuracy specification such as "accuracy required $\pm$1%". The values given in the table below represents repeatability at a two sigma confidence level; this means that if a feeder is claimed to be 1% accurate at a feed rate of 6,000 lbs./hr. and 30 or more consecutive one minute samples are taken, then 95% of the samples will weigh between 99 and 101 lbs., assuming perfect linearity.

	Gravimetric	Volumetric
Good Flowing Materials	0.2 to 0.5%	1 to 2%
Poor Flowing Materials	0.5 to 1.0%	worse than 2%

As a rough guideline gravimetric feeding will improve accuracy over volumetric feeding by a factor of about 5, but again this number may vary widely depending on the application and ingredient fed. For example, take a material that is good flowing, non-packing and has a constant bulk density independent of material head. If this material can be fed a 1% volumetrically, then it is reasonable to expect gravimetric accuracies in the 0.2 to 0.3% range. On the other hand, it is highly unlikely that a material with wide bulk density variations and/or erratic screw filling can be fed even gravimetrically at much better than 1% due to the wider variations in the control signal to the variable speed drive of the feed mechanism.

Volumetric accuracy is essentially a function of the ability of the feeder to maintain constant the ingredient bulk density and the feeder speed. As mentioned previously feeder manufacturers employ a wide variety of mechanical techniques in order to condition the ingredient into a constant density in the feed mechanism and they all have their relative advantages and disadvantages. Speed regulation can be improved by the addition of a tachometer to provide a positive speed feedback signal allowing closed loop control of motor speed.

Gravimetric accuracy on the other hand is affected by additional factors such as scale design, load cell precision, control algorithm, etc., which will not be discussed in detail here. We can sum these up by stating that a gravimetric feeder compensates for changes in such areas as material density, screw filling, humidity and temperature, while a volumetric feeder does not. This, in essence, is the reason for the accuracy improvement.

The most common applications for gravimetric feeders are where raw material costs are high or final product quality control is stringent. In these cases the initial price premium for a gravimetric feeder can be paid back over a relatively short time period when compared with wastage of expensive raw materials or frequent rejection of off spec final product. While the user must define his own process requirements, gravimetric feeding should seriously be considered when accuracy required is better than 2%.

b) Information Availability - a gravimetric feeder makes available such information as instantaneous feed rate, alarm history, totalized flow over a selectable time period and weight of material in the weigh hopper or on the weigh deck. This information is available either at the feeder controller or can be transmitted over a serial link for use as required by external control devices (e.g. PLC's, host computers, etc.). A volumetric feeder cannot provide this data, so the user must carefully consider the importance of data availability prior to specifying a particular feeder type. For example, gravimetric feeders are quite often specified in cases where accuracy is not of importance (5% or so being acceptable but the use requires a totalized flow for inventory or statistical process control purposes.

c) Fault Detection - a gravimetric device includes built in alarms to notify the operator of a process upset condition. Consider the situation where a bin located over a weigh belt feeder runs out of material. The weigh belt feeder will quickly alarm, either on low belt load or high motor speed, shut the process down and alarm the operator for corrective action. A volumetric belt on the other hand will merrily continue to operate at its preselected motor speed requiring the fault to be detected by other means, with perhaps a much larger time delay intolerable to the process.

d) Continuous Mixture Control - the most common application here is where 2 or more ingredients are required to be fed into a continuous mixer or extruder. The control of the ingredients may be independent (where the setpoint of each feeder is set individually or dependent (when slave feeders are slaved to a master feeder or master rate setter). Gravimetric feeders are needed whenever the mixture control must be done by weight, usually for reasons of accuracy.

e) Price - in a Utopian world with unlimited budgets, the volumetric feeder would probably disappear with only gravimetrics being specified due to their much wider scope of operation. The flip side of the coin of course if price; the required addition of controllers, load cells, scales and ancillary electronic equipment makes a gravimetric feeder approx. 2-5 times as expensive as the corresponding volumetric unit. Feeders should never be purchased based on price alone. A feeder is an important investment with millions of pounds of material passing over it in its lifetime. For this reason the user must make a careful payback analysis based on process requirements and budgetary constraints in an effort to obtain the maximum return on his investment.

Step 3 - Weigh Belt vs. Loss In Weight

Assuming that the user has specified a gravimetric feeder for his requirement, we will now examine the decision process between the 2 most common gravimetric feeders, weigh belts and loss in weights. The comparison is summarized in Table 1 below.

TABLE 1

Selection Comparison Between Weigh Belt Feeder vs. Loss In Weight Feeder

Selection Criteria	Weigh Belt Feeder	LIW Feeder
a) Self taring	no	yes
b) Component accessibility	poor	good
c) Type of materials fed	narrow range (wider with pre feeder)	wide range
d) Dust containment	poor	good
e) Feed rate	higher rates	lower rates
f) Accuracy	depends on system parameters	
g) Use as meter	yes	no
h) Headroom required	low	high
i) Refill required	no	yes
j) System pressure	higher pressures possible	max. 6" W.C.
k) Hazardous area classification	easier	more difficult
l) Price	usually lower	usually higher

A few of these are examined in more detail below:

a) Component Accessibility

The weight sensing device in a weigh belt feeder is generally located in or close too the product zone, although in some applications it may be possible to cantilever it outside. This poses special design considerations for handling dusty or hot product streams.

A LIW has no electronic or drive components in the product stream increasing the design flexibility for the above applications. In a suspended type LIW scale the load cell, levers, fulcrum and mass counterbalance are all located external to the feeder in a very accessible position for servicing if required. This is more difficult in a platform scale since the entire feeder has to be lifted to access the electronic and scale components making service more costly and time consuming.

b) <u>Dust Containment</u>

This is a major problem area in improperly installed weigh belt feeders. While belt feeders do come equipped with top and side calibration and may get misplaced in the course of time resulting in a pile of material building up under the feeder. There are a few ways to minimize weigh belt problems with dusty materials:

- install full bottom dust collection hopper under feeder with connection for exhausting dust through central dust collecting system;

- install dust connection right at feeder discharge where the major dusting problem occurs;

- install pre-feeder to lay the ingredient in an orderly fashion onto belt and minimize dusting;

- in some cases, it may be possible to install an oversized belt and utilize only the middle section for actual conveying of ingredient.

While the above suggestions will improve belt feeder performance, they may not eliminate the dusting situation completely. LIW feeders do not have this problem as they are completely sealed so they enjoy the advantage when housekeeping or employee safety (i.e. when handling toxic ingredients) are issues under consideration.

c) <u>Feed Rate</u>

It is impossible to state a number which one should use one type of feeder and above the other as there are obviously many other factors to be considered. What we can say is that the higher the feedrate is, the more attractive weigh belt feeders become compared with LIW's, especially when evaluating price, headroom and refill requirements.

The following comparison is a general summary of feed rate vs. feed type, keeping in mind that the numbers will vary among manufacturers:

Feed Rate (lbs./hr.)	Comments
under 60	Belt feeders not used. LIW's good down to 0.5 lbs./hr.
60 - 1,000	Both types OK, with perhaps a slight edge to LIW's.
1,000 - 10,000	Again both types OK, but advantage now shifting to belts.
over 10,000	LIW's relatively rare, but sometimes required due to nature of material. Belt feeders common and often use pre-feeders.

d) Headroom

A big advantage for weigh belt feeders here. Whereas all weigh belts from 100 to 200,000 lbs./hr. require maybe 2-3 feet of headroom (not including pre-feeders, if applicable), LIW feeders quickly get very large as the capacity increases, with a 50,000 lb./hr. unit requiring 12-13 feet or more. In addition, there is also the refill device to consider in the LIW case. IN situations where the headroom simply cannot be increased without major expenditures this does result in weigh belts being used where a LIW might be a more optimal choice. LIW manufacturers can help by increasing the refill frequency which decreases the weigh hopper size but there is a limit to this especially in the case of high displacement scale designs which take a long time to stabilize and return to gravimetric control after refill.

e) Hazardous Area Classification

Weigh belt feeder accuracy is not affected due to the requirement for explosion proof motors which are much heavier than the comparable TEFC/TENV motors. The effect on the feeder depends on the scale design. Suspended and counterbalanced scales are affected only slightly since the extra dead weight is counterbalanced out. Non-counterbalanced scales suffer a big loss in resolution and accuracy since a larger portion of the available weighing range is used to weigh the dead load and not the material. Platform scales may suffer further loss of stability due to possible overhang of the large motors.

f) Price

Weigh belt feeders without prefeeds in mild steel construction range in price from $20,000 to $35,000. Loss in weight feeders in mild steel construction range in price from $25,000 to $50,000 for feed rates up to 5,000 lbs./hr.

Price of course should be the last selection criterion and be employed only once all other relevant areas have been examined. A feeder cheaper in initial cost but misapplied will undoubtedly prove to be more expensive in the long run.

6.0 Conclusion

Feeders form a critical components in many metallurgical processes. Feed rate accuracy and feeder reliability greatly affect end product quality. The guidelines presented in this paper will assist the user in the feeder selection process, but keep in mind they are guidelines only. Feeding is as much an art as it is a science; while on some applications a clear favourite stands out, on others several process solutions may be feasible. The educated buyer, by combining a judicious application of the selection parameters along with his own plant experience and the recommendations of the feeder manufacturer, can greatly increase his chances for a successful and profitable installation.

7.0 Bibliography

1. A. Kovats, "Compared Features and Design Considerations When Selecting a Volumetric Feeder, Weigh Belt Feeder or Loss In Weight Feeder" - Powder & Bulk Solids Conference, May 15-18, 1989.

2. T. Fahlenbock, " The Principles of Dynamic Weighing Applied to Chemical Feeding Equipment of the Loss In Weight Type" - Powder & Bulk Solids Conference, May 12-14, 1987

FIGURE 1

COMPONENTS OF A CONTINUOUS LOSS–IN–WEIGHT FEEDER

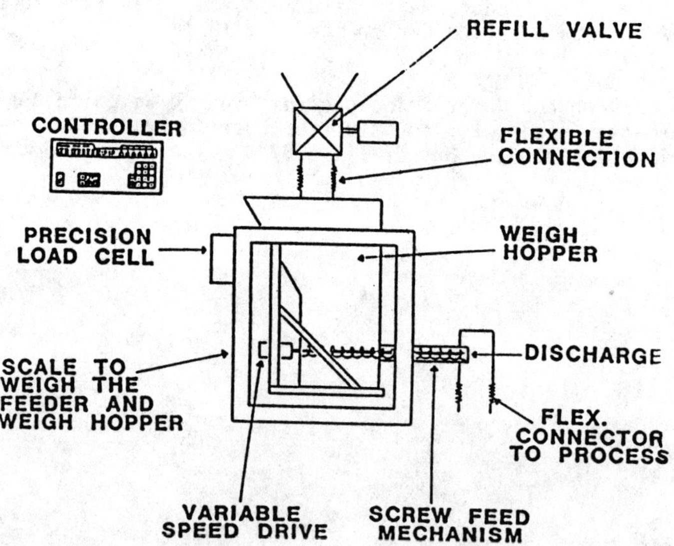

What's new in continuous weighing

G. Rebucci
Schenck Weighing Systems, Fairfield, New Jersey, U.S.A

I. Introduction

Continuous Gravimetric Weighing Systems for bulk solids were first introduced at the turn of the century to provide a means of weighing material as it flowed on a conveyor belt. This invention led to the development of the Weighbelt Feeder, which is a conveyor system especially designed for weighing. The Weighbelt Feeder served as the primary device in the continuous weighing industry for many years and was an essential part of the conversion of numerous processes from discontinuous or batching systems to continuous systems. The first machines were mechanical devices and were highly developed over a long period by several companies. Mechanical designs dominated the market well into the sixties, at which time the strain gauge load cell reached acceptable levels of performance and reliability and the industry entered the analog instrumentation generation. Electronics opened up the industry to more companies and development accellerated through digital electronics, hybrid systems and into the present microprocessor based technology.

Today, systems are significantly enhanced by the use of strain gauge load cells and microprocessor based electronics. Both of these concepts are well developed and provide the possibility for precise force measurement, accurate data processing, flexible control schemes and very high reliability. Weighbelts are still a large part of the continuous weighing market, however there are presently some new concepts which have found increasing application.

II. New Technology in Continuous Weighing

a) The Solids Flowmeter

The name Flowmeter clearly establishes that this device is primarily a meter and its basic function is to measure the flow rate of bulk solids. This is directly analogous to the function of a liquid flowmeter. The Solids Flowmeter is shown in Figure I. Material is fed into the Guide Chute and flows along this chute into the Flowmeter housing. Inside the housing, the material flows from the Guide Chute onto a special "Curved Measuring Plate." The Curved Measuring Plate" is mounted on a lever system such that the vertical resultant force is sensed by a load cell. Horizontal forces are effectively isolated from the load cell to allow the use of horizontal vibration of the measuring plate. Vibration is optional and only required for stickier material.

The standard Flowmeter is inherently very well sealed and is dust tight and suitable for inert gas blanketing. The Flowmeter can be used in very hot applications. However, if the temperature exceeds 200 deg. f. then the sealing is compromised.

The operating principle is illustrated in Figure II. The material is fed into the Guide Chute and slides into the Flowmeter and around the "Curved Measuring Plate." The material enters the measuring plate at velocity V1 and leaves with velocity V2. The magnitudes of V1 and V2 are essentially equal, the force results from the change in direction. It is important that V remain constant so the introduction of material into the Guide Chute is very important. In general this means that the feeding system into the Guide Chute must introduce the material horizontally or in other words establish a consistent feed point regarding the initial velocity of the material. The lever system is very important because the pivot point is located on the tangent of the Measuring curve in order to cancel any friction force as the material flows across the plate. The lever system includes a set of special flexures to isolate the vertical and horizontal resultants of the material force in order to enable the use of the Measuring Chute Vibration as previously described. The force on the measuring plate is a function of the Mass Flow Rate (M, kg/hr) and the change in velocity (V=V1-V2). Force = (M,V)

b) Loss-in-Weight Feeders

The principle of continuous weigh feeding by Loss-in-Weight has been developed to a prominent position in the industry. It may be stated that this principle is inherently superior technologically but requires very high electronic and mechanical sophistication to realize operating systems.

A Loss-in-Weight feeder is shown in Figure III and consists of a Weigh Hopper, Feed Section and Feed Screw. The Weigh Hopper is sized to contain a volume of material equal to 10-15 minutes supply at the maximum system feed rate.

For Example:

If the feeder is to deliver a material with a density of 1.0 ton/cu. M at a maximum capacity of 4000 kg/hr then the required weigh hopper volume is calculated as follows:

$$\text{Volumetric Capacity} = \frac{4000 \text{ kg/hr}}{1.0 \text{ ton/cu.M} \times 1000} = 4 \text{ cu.M/hr}$$

$$= .0667 \text{ cu.M/min}$$

Weigh Hopper Capacity = .0667 x 12 = .8 cu. M
(for 12 minutes capacity)

The feed section must be carefully designed along with the Weigh Hopper to properly suit the handling characteristics of the material. FIG. III shows a typical Loss-in-Weight feeder including the Weigh hopper and Feed Section supported on Load Cells. FIG. III shows a horizontal agitator. The agitator rotates at a fixed speed in the range of 5-12 rpm and functions to deaerate and stabilize the density of the material in the Feed Section. A horizontal agitator is suitable for free flowing, fine, dry powders. For cohesive and sticky materials, the design must have a round Weigh Hopper with a vertical agitator. The feed section in FIG. III is further equipped with a variable speed feed screw which is also carefully selected based on material handling properties. Finally, all the above are supported on load cells.

FIG IV provides the characteristic equations describing the Loss-in-Weight principle including differentiation of the load cell signal to generate feed rate.

FIG IV illustrates the operating sequence for the Loss-in-Weight Feeding System.

The Weigh Hopper is intially filled with material and as the feeding section operates, the load cells will sense the decreasing weight of material. This signal is used by the Microprocessor to calculate the actual feed rate (kg/min) from the system as well as to totalize the amount of material fed.

The Microprocessor also compares the actual feed rate to the set point and sends a signal to the drive for the feed screw to increase or decrease the RPM such that the actual feed rate being measured equals the set point. When the material in the hopper reaches a predetermined low level, the Microprocessor initiates a fill cycle. The feed screw is switched to a fixed volumetric speed and the fill cycle may now be initiated as shown in FIG V. The fill cycle is critical to the overall operation of the Loss-in-Weight system because consistent filling must be achieved in a relatively short time so that the ratio of gravimetric feeding time to volumetric feeding time is high. The fill cycle sequence is as follows:

1. Silo Conditioning: FIG V shows the use of a pulse aeration system which is used for free flowing powders. Aeration nozzles are installed on the bottom cone of the silo in two or three circular arrays. Solenoid valves are actuated by the Microprocessor system typically in two half second bursts; one or two seconds apart. This slightly aerates the material in this section.

2. Vent Valve Open: Venting is required to allow the weigh hopper to fill quickly and to prevent excessive dusting of the area. Note that during normal operation the flexible connection between the weigh hopper and the silo must be at atmospheric pressure otherwise this will act as a diaphram and introduce error forces onto the load cells.

3. Fill Valve Open: The fill valve is usually a pneumatically operated butterfly type.

4. Fill Valve Closed: The microprocessor utilizes the load cell signal to close the fill valve when the predetermined amount of material is in the weigh hopper.

5. Blow Down Cycle: A metering cone is installed directly below the fill valve and serves to reduce and control the flow rate of material during refilling. After the fill valve closes, an air jet system is activated by the microprocessor to introduce air to clear the dust from the system.

6. Vent Valve Closed: Once the blow down is completed, the vent valve may be closed without danger of further dust emission from the system.

The fill cycle is now complete and the system may return to gravimetric operation. The entire fill cycle is controlled by the microprocessor and should require between 8 and 12 seconds.

The time between fill cycles should be 3 minutes or more to provide a Gravimetric/Volumetric ratio of at least 15.

FIG. VI illustrates important considerations of material handling required for a successful Loss-in-Weight system for fine free flowing powders. The first consideration is the design of the weigh bin with steep sides and polished stainless construction to promote mass flow. The fill cycle introduces the new material in an aerated condition and the material must be contained and held away from the feed section so it will deaerate. The buffer amount is important and is normally in the range of 4 to 6 minutes supply. Note that the Agitator is always entirely inside the Buffer area and functions to greatly increase de-aeration prior to the material entering the Feed Screw. The Feed Screw is also of special design and includes a Mass Flow design under the Weigh Hopper. The delivery section includes a compaction section to form a plug followed by an expanding section to reaerate the material for consistant discharge. The plug section provides a safeguard against flooding and is required on very difficult to handle materials. The expanding section is always important to smooth out discharge flow pulsations which occur with any screw. The material should discharge in a smooth uniform stream.

This describes the basic operation characteristics and requirements for a Loss-in-Weight feeder for fine powders. Other materials are similarly handled using appropriate Feed Sections.

c) The Mass Flowmeter

The utilization of particle acceleration to measure mass is a simple, direct and well estabilished method. This technology is well known for measuring the flow rate of liquids using Coriolis Meters and can now be applied to bulk solids. This new device provides two important benefits for continuous weighing processes:

- Guaranteed accuracy to material without material testing.

- Long term calibration stability.

A Mass Flowmeter is shown in FIG. VII. It consists of an inner and outer sheet metal housing. The inner housing holds an AC motor mounted on bearings with a load cell/torque arm system. The motor shaft extends vertically upward through a frictionless seal into the outer housing. The motor shaft is fitted with an impeller wheel with a cone in the center and radial guide vanes on the perimeter.

The Flowmeter must be fed material from some type of volumetric prefeed device located above. The material may free fall up to 6 ft or so onto the central cone. The material flows horizontally across the rotating wheel through the guide vanes which impart a rotation or Coriolis acceleration to the material. This acceleration results in a torque reaction which is directly proportional to the mass flow rate through the Flow Meter. The motor is also equipped with a high resolution digital tachometer which provides a measurement of motor rpm. The load cell and speed measurements are monitored by a microprocessor based measuring system which calculates the material flow rate (TPH) and also totalizes the throughput (tons).

FIG. VIII and FIG. IX are cutaway illustrations showing the basic configuration and material flow path for the Mass Flowmeter.

A positive ventilation system works in conjunction with a temperature monitoring system to ensure reliable operation and long life for the motor.

Calibration

The Flowmeter provides a unique feature wherein the span calibration is entirely predictable and does not require a calibration device or material sample test.

Accuracy

Accuracy of ± 1% of actual flow rate over a 5 to 1 operating range.

In the event that a material test can be done, the accuracy can be improved to ± 1/2%.

Application

Free flowing powders or granules in the range of 100 mesh to 3/16" and the flow rate range of 1 to 60 TPH.

The Schenck Flowmeter can be used as a rate control system by closing the loop to the prefeed device. This device can be used in all continuous weighing applications including measuring, rate control, batching and blending applications.

FIG. X includes a force diagram to explain the operation and characteristic equations. The centrifugal force and friction forces are radial and thereforce have no effect on motor torque. Only the Coriolis Force is transmitted to the load cell and provides the relationship.

$$\text{Force } F_c = 2\ mWV_c$$

Where

m = Mass Flow Rate
W = Radial Velocity
Vc = Radial Location

Overall torque resultant on the motor is:

$$M = m\ W\ R2$$

Note that since W is included a speed sensor is required to optimize instantaneous accuracy measurement.

$$[\text{Mass Flow Rate}] = \frac{\text{Torque (M)}}{W\ R2}$$

Conclusion:

New technology provides increased possibilities for commercialization of continuous processes with attendant advantages over batch processes. FIG. XI provides a tabulated comparison of the application ranges and suitability of these new devices.

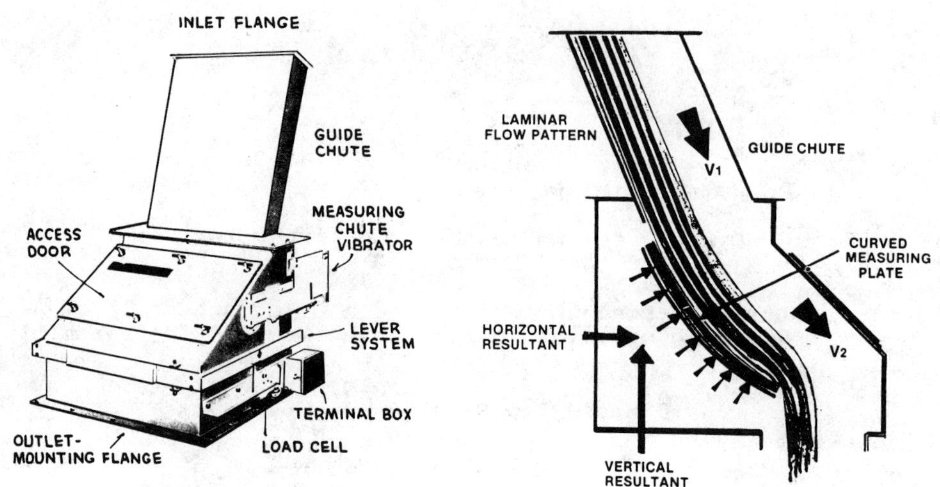

FIG. I - THE SOLIDS FLOWMETER. FIG. II - SOLIDS FLOWMETERS USING THE MOMENTUM OR CURVED PLATE MEASURING PRINCIPLE.

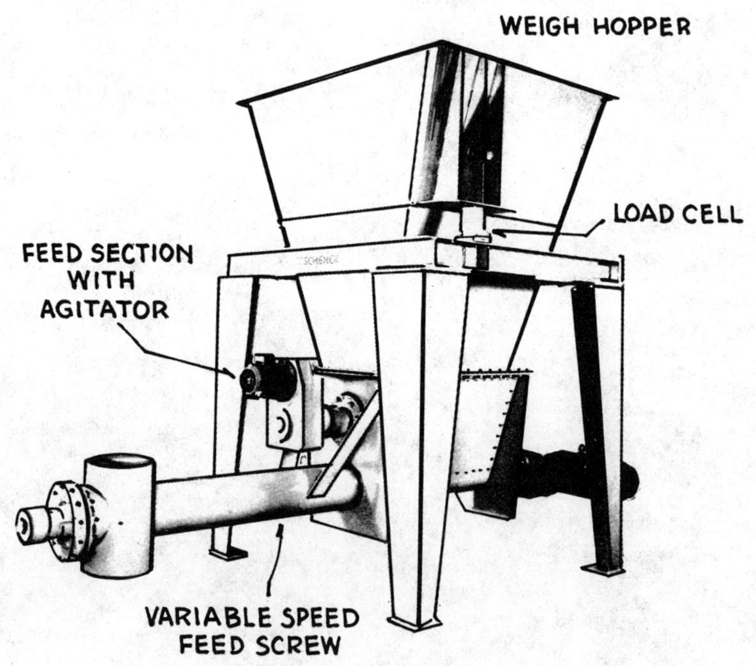

FIG. III - THE LOSS-IN-WEIGHT FEEDER

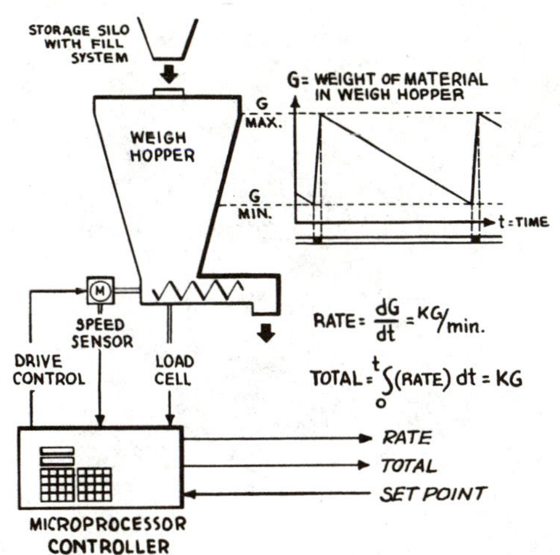

FIG. IV - LOSS-IN-WEIGHT

PRINCIPLE OF OPERATION

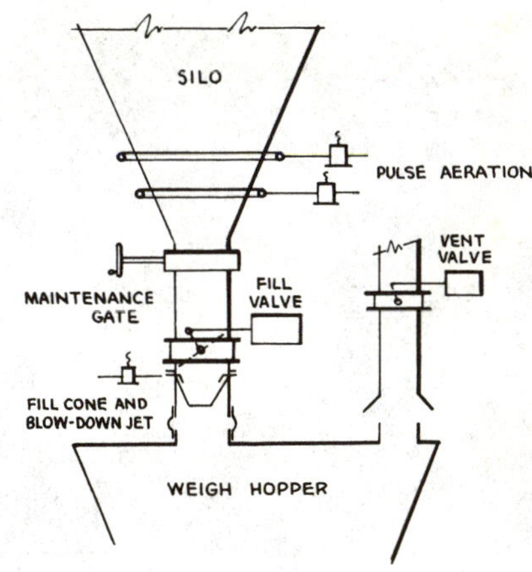

FIG. V - FILLING AND VENTING

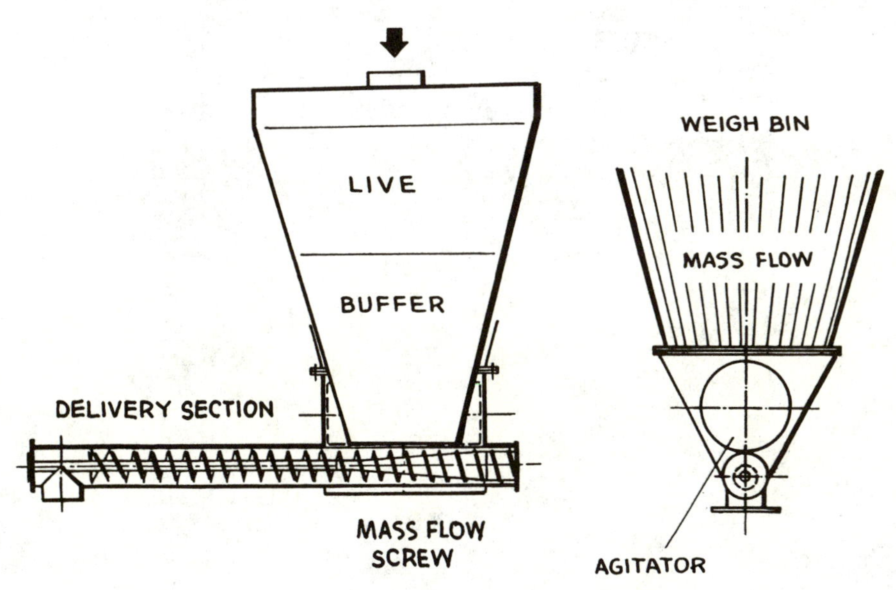

FIG. VI - MATERIALS HANDLING TECHNIQUES .... FINE POWDERS

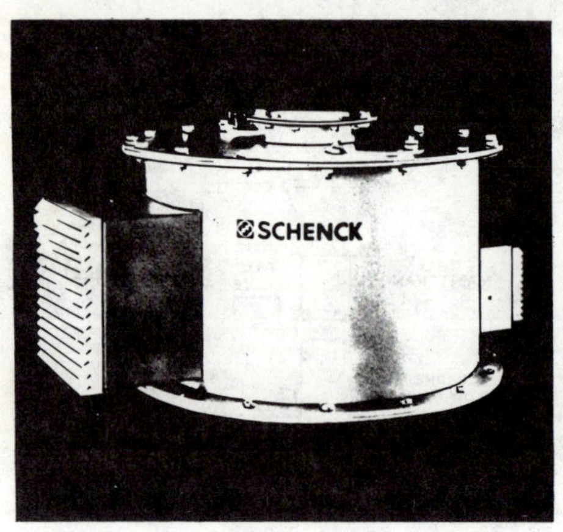

FIG. VII - MASS FLOW MEASUREMENT FOR CONTINUOUS WEIGHING OF BULK SOLIDS

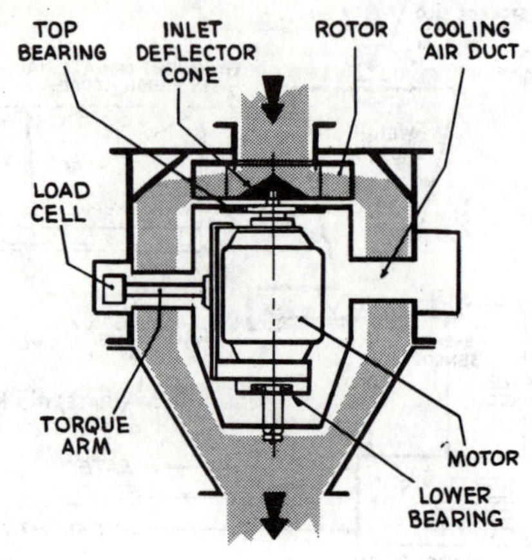

FIG. VIII - MASS FLOWMETER

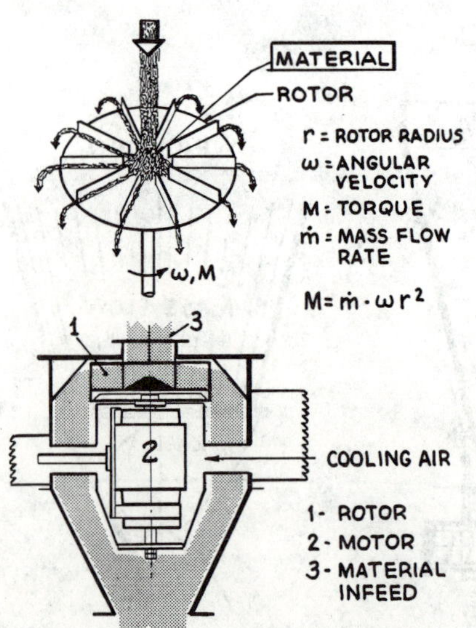

FIG. IX - MASS FLOWMETER CORIOLIS PRINCIPLE

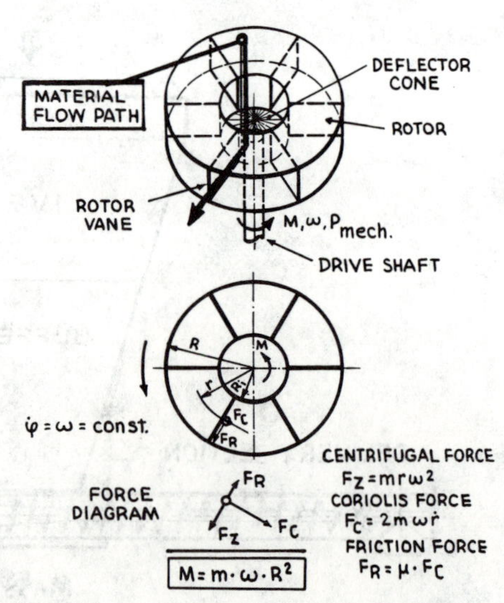

FIG. X - MASS FLOWMETER PRINCIPLE OF OPERATION

TECHNICAL CRITERIA	FLOW METER	MASS METER	LOSS-IN-WEIGHT
● MATERIALS HANDLING			
MATERIAL:			
Free-flowing	++	++	++
Bridging	O	+	+
Flushing	++	+	++
Sticky	–	–	O
Entangling, matting	O	O	+
Abrasive	+	+	+
Occasionally lumpy	–	–	–
MATERIAL TEMP. HIGH:	++	O	+
LOW:	+	+	+
● FLOW RATES AND ACCURACIES			
NOMINAL: Min. approx.	2 TPH	1 TPH	50 GM/HR
(at 60 lbs./cu. ft.) Max. approx.	1000 TPH	60 TPH	100 TPH
RANGE OF MEASUREMENT:	5:1	5:1	10:1
ACCURACY:	± 2%	± 1%	± ½ %
● INSTALLATION CHARACTERISTICS			
Headroom requirements	O	+	O
Vibration and shock	+	+	+
Wind and weather	+	+	O
Dusty environments	++	+	++
Available enclosures:			
Dust-tight	++	++	++
Gas-tight	++	++	++

++ VERY WELL SUITED
\+ WELL SUITED
O CONDITIONALLY SUITED
– UNSUITED

"the **SCHENCK** *option"*

FIG. XI - PRODUCT LINE SELECTION CRITERIA
With comparative evaluations for Flowmeters, Mass Meters, Loss-in-Weight

On-line mass flow rate measurement for pneumatic injection systems used in metallurgical processes

S.R. Woodhead, R.N. Barnes and A.R. Reed
The Wolfson Centre for Bulk Solids Handling Technology, School of Engineering, Thames Polytechnic, London, U.K.

Summary

This paper reviews the principles of a number of techniques applicable to On-Line Mass Flow Measurement for Pneumatic Injection Systems, and indeed pneumatic conveying systems in general.

It has been established for some years now, that two independent measurements are required in order to obtain the product mass flow rate in pneumatic conveying pipelines. The first measurement is that of the average velocity of the particles being covered through the pipeline. The second is that of the concentration of product in the pipeline. These two values can then be combined to obtain the required product mass flow rate.

The techniques reviewed in this paper are:

For concentration measurement,

- i) Capacitive sensors
- ii) Electrostatic sensors
- iii) Flow modulated ionising radiation

For velocity measurement,

- i) Cross correlation of axially spaced sensors
- ii) An estimated particle slip technique.

1. Introduction

In recent years there has been an increased demand in the electrical power, cement, and steel industries for greater energy efficiency. At the same time there has been an increased dependency on plant automation in these and other industries. Throughout all of these types of plant, the pneumatic transport of powdered and granular materials through pipelines is commonplace. For example, in power generation, pulverised coal is fed to the boilers by pneumatic means. In the steel and cement industries, pulverised or granular coal is also injected into the blast furnaces and kilns pneumatically. Pneumatic conveying of powdered and granular materials is also used extensively in other industries.

Accordingly, some years ago a need was identified for an on-line mass flow rate measurement technique, which could provide a reliable real time value for the amount of product being transported through a pipeline at any given time. Such a technique would enable much more efficient direct automatic control of boilers, kilns and blast furnaces which are recognised as being major problem areas in terms of automatic control. The technique would also be invaluable to operators of process plant, such as with batch mixing systems, where the masses of each constituent ingredient could be evaluated as the process proceeds. Provided the mass flow rate measurement technique is sufficiently accurate, this would then eliminate the need to install costly and sometimes less than accurate load cell and belt weighing systems.

Initial efforts to provide such a technique were focused on adaptations of techniques originally devised for measuring the flow rate of single phase flows. For example, orifice plates and venturi meters (Ref. 1). However, for various reasons, these proved to be less than satisfactory. Realising that this approach was fundamentally flawed, movement was made towards the development of techniques specifically for two phase flow applications. Almost all of these new techniques rely upon the independent evaluation of the average particle velocity and the concentration of product in the pipeline. The two values, together with a calibration factor are, then multiplied to obtain the product mass flow rate. Many of these specialised systems are still the subject of research or on-going development. Indeed, there is, to the authors' knowledge, only one system of this type in use in industry, and it is known that the conditions under which this will operate satisfactorily are somewhat limited.

In this paper the authors will initially examine the possible techniques for the measurement of concentration. The principles of the velocity measurement techniques will then be outlined.

Most of this material will be take from work which is either on going or is scheduled for investigation in the very near future at The Wolfson Centre for Bulk Solids Handling Technology, Thames Polytechnic, London, UK.

2. Concentration Measurement Techniques

There are currently three possible techniques applicable to concentration measurement under investigation. The most established technique will be dealt with first and then attention will move on to those approaches which are less well established.

2.1 Capacitive Techniques (Ref. 2-12)

This approach involves the use of insulated sections of pipeline. The capacitance between the insulated section and the rest of the pipeline is then measured by incorporating the capacitor in an oscillator circuit. As particulate product is introduced into the pipeline the dielectric of the capacitor will change, thus causing the capacitor value, and hence the oscillator frequency to change. Depending on the flow conditions, two different configurations of transducers are in use, Figure 1.

The transducer electronics are arranged as shown in Figure 2. The oscillator output is fed to an F.M. demodulator, which provides a voltage output as a function of capacitance. In order to keep the F.M. demodulator within the linear region, feedback is applied by using a low pass filtered version of the output to feed a varicap diode in the oscillator circuit. An A.C. amplifier is then added to the F.M. demodulator output to provide the transducer output. The output from the transducer will then constitute an A.C. signal. It has been found that the amplitude of the A.C. signal is a function of the product concentration in the pipeline.

The range of product to air mass flow ratios over which these systems will operate is somewhat limited and one manufacturer (Ref. 2) does not recommend their use in systems operating at product to air mass flow ratios below the values of 2 to 7 - depending upon the particular product under consideration. This is due to the fact that the change in the dielectric becomes very small at the low solids loadings. Unfortunately, many pneumatic conveying systems operate below these values - particularly in the power, cement and steel industries, where there is a considerable need for such a system. Thus, to date applications have been somewhat limited. It is expected that within the next few months a system of this type will undergo extensive trials at The Wolfson Centre in a specially commissioned test plant. It is the object of this work to determine the exact limits of applicability of this technique for coal and other materials.

2.2 Electrostatic Techniques (Ref. 13-19)

This technique involves the use of pipeline sections similar to those already shown at the bottom of Figure 1. However, rather than incorporating the sensor in an oscillator circuit, as in the capacitive transducer, the insulated ring section is connected to the input of a charge amplifier. As the particles pass along the pipeline they will cause the value of the electrostatic charge on the ring sensor to vary. Once again a noise signal is obtained and the R.M.S. value of this noise is a function of the concentration of product in the pipeline.

Initial tests indicate that this type of transducer may operate more effectively at the low product to air mass flow rate ratios encountered in the power generation and other industries. However, a full test program is not yet complete and more reliable information will be available in the near future.

2.3 Flow Modulated Radiation Techniques (Ref. 20)

The basic principle of this technique is shown in Figure 3. Radiation, either, Gamma, X-ray or Beta particle is introduced into the pipeline via a window, The intensity of the radiation is then measured on the opposite side of the pipeline, usually with a scintillation counter. It can be seen from Figure 3 that only a small part of the pipeline cross section will be interrogated by the radiation form the single source. This is the main weakness of this technique, since the distribution of the product within the pipe cross section is rarely uniform, because of the 'roping' effect (Ref. 21). Consequently, movement is now being made towards the use of multiple sources and counters, in an attempt to interrogate a larger proportion of the pipeline cross section. The basic principle of operation is very simple. As product is introduced into the pipeline, the radiation will undergo attenuation across the pipeline. The attenuation will be a function of the product concentration.

The research and development being carried out into this technique is based in a number of areas:

i) The window material - in order to minimise the attenuation caused by the window itself.

ii) The types and the configuration of the sources and counters in order to minimise both cost and radiation hazards, and maximise the proportion of the cross section being interrogated.

iii) The processing of the signal produced by the counter(s) - the laws governing the attenuation of ionising radiation suggest that the process is not linear and some signal processing is needed. This problem becomes particularly acute with multiple sources and counters.

Although a small amount of work has already been undertaken to explore the possibilities of this technique, clearly, there is still much to be done. This technique is therefore one which the authors shall be continuing to examine in some detail.

3. Velocity Measurement Techniques

As previously mentioned, evaluation of both the product concentration and the average particle velocity is necessary in order to obtain the mass flow rate of product in the conveying pipeline. Two techniques applicable to the measurement of the average particle velocity will now be discussed. Again, the most established technique first, which will be followed with a newer, lesser known method.

3.1 Cross Correlation of Axially Spaced Concentration Sensors

The principle of this technique is shown diagrammatically in Figure 4. Two concentration sensors are spaced axially along the pipeline, and assuming that the flow pattern remains constant along the short length between the sensors, two signals will be obtained as shown. It should be noted that, as we would expect the signals are similar, and time delayed by period T. By obtaining the cross correlation function of the two signals in real time, it is possible to evaluate the delay period and hence, (assuming that the sensor separation is known) the average particle velocity. In principle, it should be possible to employ any of the concentration measurement techniques in the axially spaced sensors. Both the capacitive and electrostatic techniques have been used with some degree of success. However, once again the results of exhaustive tests are not yet available. To the authors' knowledge, none of the radiation techniques have been combined with the cross correlation techniques, for use in pneumatic conveying applications. However, they have been utilised in other two phase flow applications (Ref. 26).

3.2 The Estimated Particle Slip Technique

This technique is based upon the assumption that the mass flow rate of the air entering the system can be measured. The principle of the technique is shown in Figure 5. If the mass flow rate of the incoming air can be determined by some conventional technique, such as an orifice plate or a critical flow nozzle, and the temperature, pressure and pipeline bore are known at a given point in the pipeline, it is possible to calculate the superficial air velocity (the air velocity with no product present).

It is known that under most conditions experienced in pneumatic conveying systems, the average particle velocity and the superficial air velocity are not the same. The average particle velocity is always less than the superficial velocity and the particles are therefore said the 'slip'. Unfortunately, the slip ratio (ratio of particle to superficial air velocity) is not constant. The value changes according to a number of factors, such as product density, particle size, conveying velocity, whether the pipeline is horizontal or vertical and the distance from the last bend or other obstruction to flow.

However, it is possible to measure the slip ratio using a specialised test facility incorporating Laser Doppler Velocimetry instrumentation (Ref. 23-25). Such a facility already exists at The Wolfson Centre and if the slip ratios for various products under various conveying conditions can be evaluated, this technique could form the basis of a mass flow rate measurement system.

If the particle velocity could be predicted with sufficient accuracy this approach could be particularly cost effective, since much existing process plant is already instrumented to give the mass flow rate of the air intake. All that would then be needed would be low cost microprocessor based signal processing system, and some means of measuring the product concentration. This technique also forms the basis of an on going research project at The Wolfson Centre.

4. Concluding Remarks

This paper has reviewed a number of techniques which it is hoped will form the basis for a reliable instrument. One particularly important application of such an instrument would be to provide a feedback signal for flow splitting control systems. Flow splitting of pneumatic conveying pipelines is used extensively in a number of industries, and controlling the splitting process is notoriously difficult. The principle of such a technique is shown in Figure 6. However, it can be seen that in order to split the flow, a control element is also needed. The development of control elements for flow splitting is the subject of another current research project at The Wolfson Centre, and the development of a complete system is a matter which the authors shall be investigating in the future.

Acknowledgments

The authors thank the following organisations for supporting the research programmes outlined in this paper:

Babcock Energy
Blue Circle Cement
The Department of Energy
Kevex Corporation (UK)
The School of Information Engineering, Teesside Polytechnic
The Science and Engineering Research Council
Tealgate Limited
The Wolfson Foundation

References

1. McVeigh, J C. Craig, R W. 'Metering of Solid Gas Mixtures using an Annular Venturi Meter', Proc. Pneumotransport 1 Conference, Cambridge, 1972.

2. Beck, M S. Wainwright, N. 'Flow Failure Detector for Powdered and Granular Materials', Control, pages 52-56 January, 1969.

3. Green, R G. Foo, S H. Mitchell, S. 'Application of Low Cost Microcomputer to Measurement and Control of Pneumatic Conveying Systems', Proc. 5th Int. Conf. Pneumatic Transport of Solids in Pipes, London, April 1980.

4. Beck, M S. 'Recent Developments and the Future of Cross Correlation Flow Meters', Proc. Int. Conf on Advances in Flow Measurement Techniques, Warwick, UK., 1981.

5. Cardon, M. Green, R G. John, R. 'Applications in Monitoring Particulate Mass Flow Rates of Abrasive Materials', Proc. Int. Conf on Advances in Flow Measurement Techniques, Warwick, UK, Sept. 1981.

6. Green, R G. Cunliffe, J M. 'On-Line Measurement of Two Phase Fluid Flow with an F M Capacitance Transducer', Proc. Int. Conf on Advances in Flow Measurement Techniques, Warwick, UK. Sept. 1981.

7. Green, R G. Foo, S H. Philips, J G. 'Flow Measurement for Optimising the Feed Rate of Pulverised Coal Fired Boilers'. Symposium on Fossil Energy

Processes, San Francisco, California, USA. 1981.

8. Thorn, R. Beck, M S. Green, R G. 'Non-Intrusive Methods of Velocity Measurement in Pneumatic Conveying', J. Physics E, Vol. 15, 1982.

9. Beck, C M. Henry, R M. Lowe, B T. Plaskowski, A. 'Instrumentations for Velocity Measurement and Control of Solids in a Pneumatic Conveyor', Proc. 1st Int. Conf. on Pneumatic Conveying Technology, Stratford-upon-Avon, UK. 1983.

10. Hammer, E A. Green, R G. 'The Spatial Filtering Effect of Capacitance Transducer Electrodes', J Physics E. Vol. 16. 1983.

11. Boeck, T 'Measurement of Velocity and Mass Flow Rate of Pneumatically Conveyed Solids by the use of the Correlation Measurement Technique', Proc. Conf. on the Flow of Particulate Solids, Bergen, Norway, August, 1986.

12. Boeck, T. 'Granucor - Mass Flow Measuring System for Pneumatically Conveyed Solids', Proc. 3rd Int. Conf. on Pneumatic Conveying Technology, Jersey, C I, March 1987.

13. Shackleton, M E. 'Electrodynamic Transducer for Gas Solids Flow Measurement', MPhil Thesis, University of Bradford, 1983.

14. Gajewski, J B. 'Charge Measurement of Dust Particles in Motion, Part II', J. Electrostatics, 15 p. 67, 1984

15. Coulthard, J. Corbett, M J. 'Metering of Pulverised and Granular Coal', Energy Transport in Pipelines, Institute of Measurement and Control, London, June, 1987.

16. Dechene, R L. Averdieck, W J. 'Triboelectricity -A New Fine Particle Measurement Parameter', 17th Annual Meeting of the Fine Particle Society, San Francisco, USA, 1986.

17. Dechene, R L. Averdieck, W J. 'Particulate Velocity Measurement by the Triboelectric Effect', Proc. of the Powder and Bulk Solids Handling and Processing Conf. Chicago, USA, 1987.

18. Dechene, R L. Averdieck, W J. 'Triboelectricity : A Parameter for Solids Flow Measurement', Powder and Bulk Engineering, June 1987.

19. Dechene, R L. ' Triboelectric Technology for Mass Flow Measurement', Proc. of the Powder and Bulk Solids Handling and Processing Conference, Chicago, USA, May 1988.

20. Howard, A V 'Development of Techniques for the Measurements of Concentration and Mass Flow Rate of Pneumatically Conveyed Coal Dust', Internal Report, CEGB NE Region.

21. Cook, D. Hurworth, N R. 'Recent Research on Pulverised Fuel Settlement in Power Station Pipelines and the Significance of "Roping"', Proc. Pneumotransport 5 Conf. London, April 1980.

22. Lech, M 'Some Aspects of Radioisotopic Measurement of Concentration and Mass Flow Rate of Powder', Proc. 5th Int. Conf. on the Pneumatic Transport of Solids in Pipes, April 1980.

23. Birchenough, A. 'The Application of Laser Measurement Techniques to the

Pneumatic Transport of Fine Particles', PhD Thesis, Thames Polytechnic, London, Sept. 1975.

24. Reithmuller, M L. Ginoux, J J. 'The Application of a Laser Doppler Velocimeter to the Velocity Measurement of Solid Particles Pneumatically Transported', Proc. of Pneumotransport 2 Conf. Guildford, Sept. 1973.

25. Birchenough, A. Mason, J S. 'An Industrial Application of the Laser Velocimeter in Gas-Solid Flows', Proc. of Pneumotransport 5 Conf. London, April 1980.

26. Byrne, B. Coulthard, J. Hampton, R. 'Comparison of Ultrasonic and Radiological Cross Correlation Measurements in Air-Water Mixtures', Proc. of the Int. Conf. on Mass Flow Rate Measurement - Direct and Indirect, London, Feb. 1989.

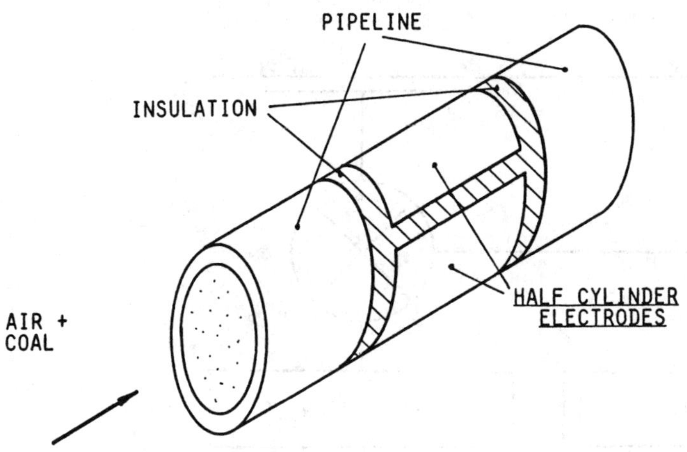

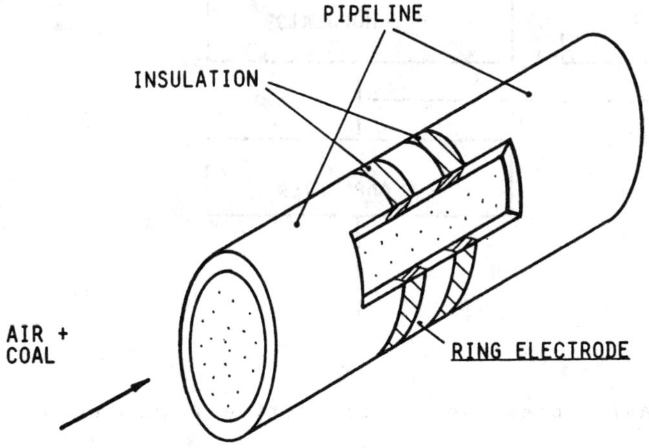

Figure 1 Electrode Configurations for Capacitive Transducers

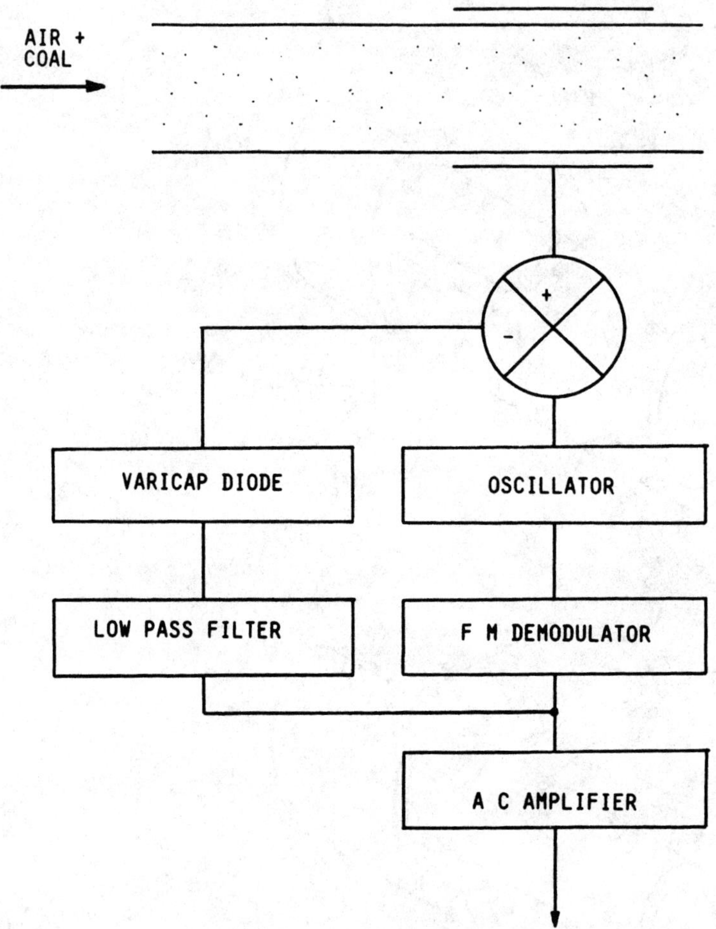

Figure 2 Signal Processing for Capacitive Transducers

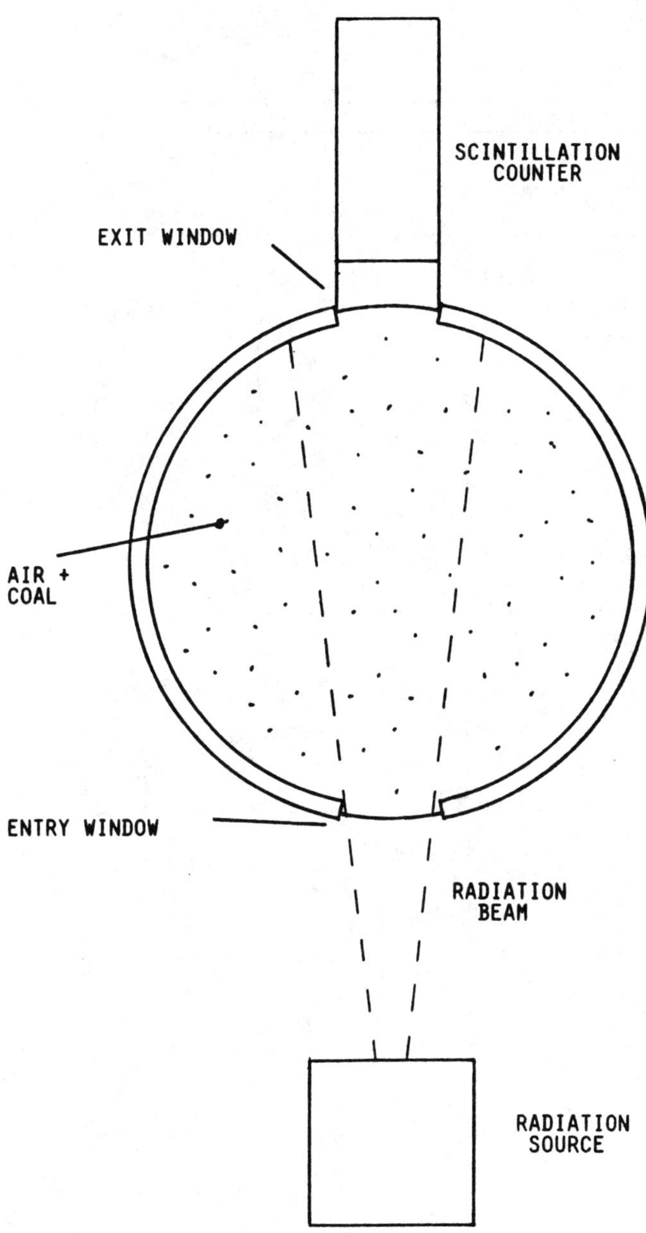

Figure 3 Principle of the Flow Modulated Radiation Technique

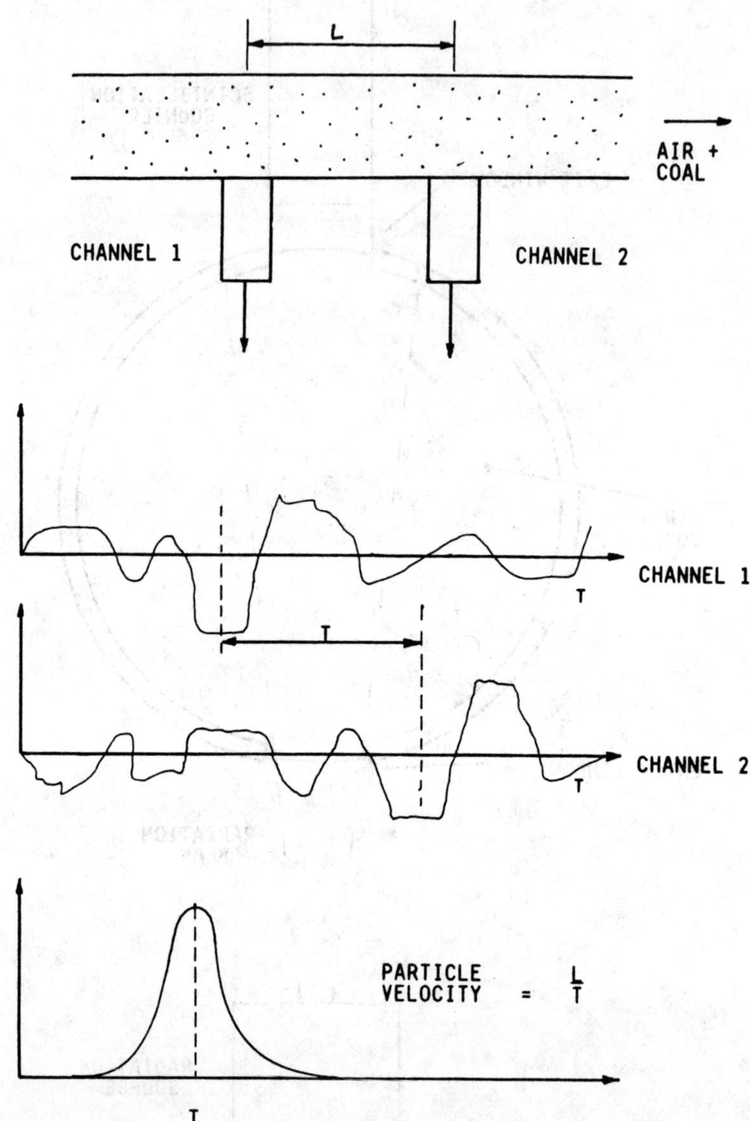

Figure 4 Average Particle Velocity Measurement Using the Cross Correlation Technique

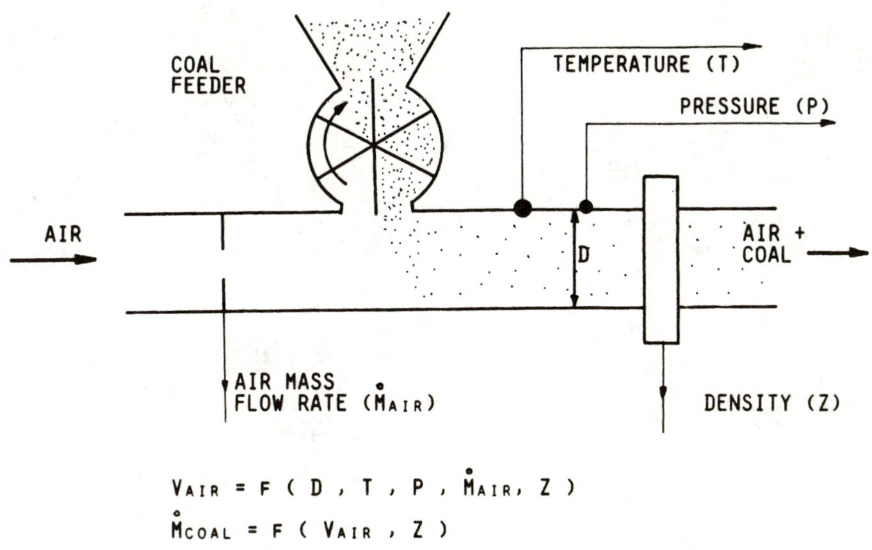

Figure 5 Principle of the Estimated Slip Technique

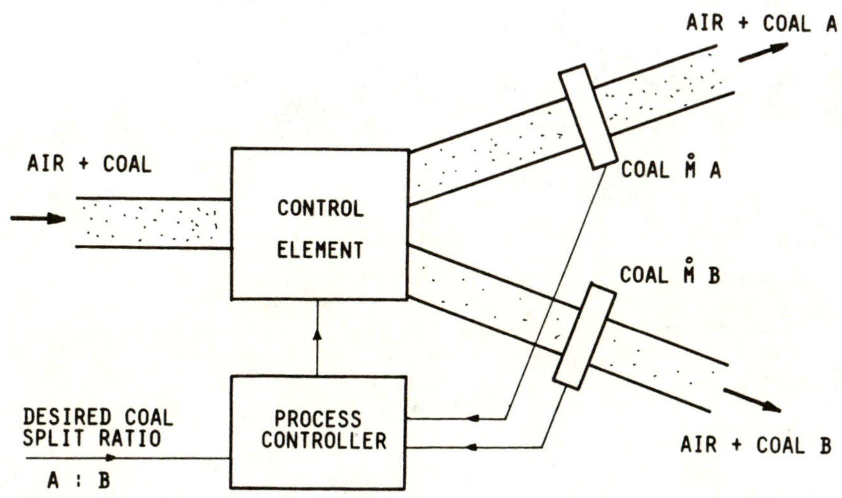

Figure 6 Principle of the Flow Splitting Control System

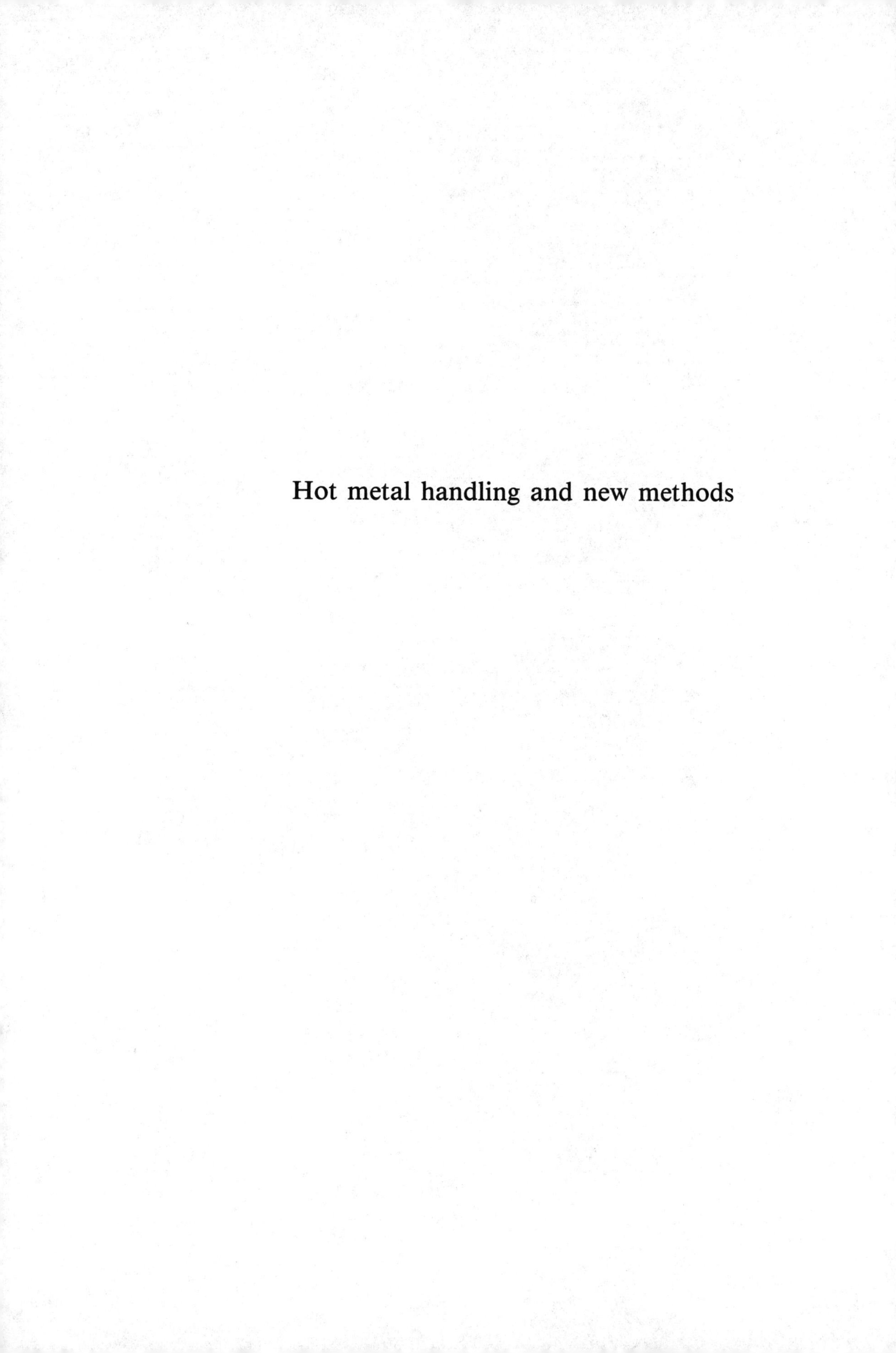

Hot metal handling and new methods

Molten metal handling — a contemporary overview

C.R. Anglin and F.P. McCaffrey
Mechanical Engineering, Hatch Associates, Ltd., Toronto, Ontario, Canada

Introduction

This paper takes a general overview of molten metal handling. A comparison of the annual tonnages of the principal metals as well as their economic significance is presented, with particular emphasis on the Canadian situation. The advances in technology or approach to the handling of each of the major metals is discussed. The handling of iron and steel is dealt with in detail because of its domination of Canadian molten metal tonnage and because handling costs can have a significant effect on the viability of operation.

Molten metal handling is considered to cover the production phase between the exit of the molten metal from the furnace and its solidification downstream. Process steps such as alloy additions may take place during this phase. These will not be dealt with in a process sense but only insofar as they impact on the handling or transportation of the metal.

Summary of Metal Tonnages and Revenues

Figure 1 shows the distribution by tonnage of the world's principal metals and their associated revenues. It is apparent that the commanding role of iron and steel exists both in quantity and in revenue. Canada's rank is thirteenth among producers. In aluminum, meanwhile, which is second in production and revenue significance, Canada is easily the third largest producer, behind the U.S. and the U.S.S.R.. Canada ranks fifth in copper, gold and silver, second in nickel and zinc, and seventh in lead.

Figure 2 is a repeat of Figure 1 but for Canadian metals. The strength of the aluminum industry coupled with its more valuable price challenges the predominance of steel to some degree. Note that the aluminum volume transported is almost one third that of steel. Nevertheless, in consideration of molten tonnage handled, iron and steel are dominant.

Since material handling represents a greater percentage of the cost of producing steel due to its lower value, the transportation of this metal will be given special attention.

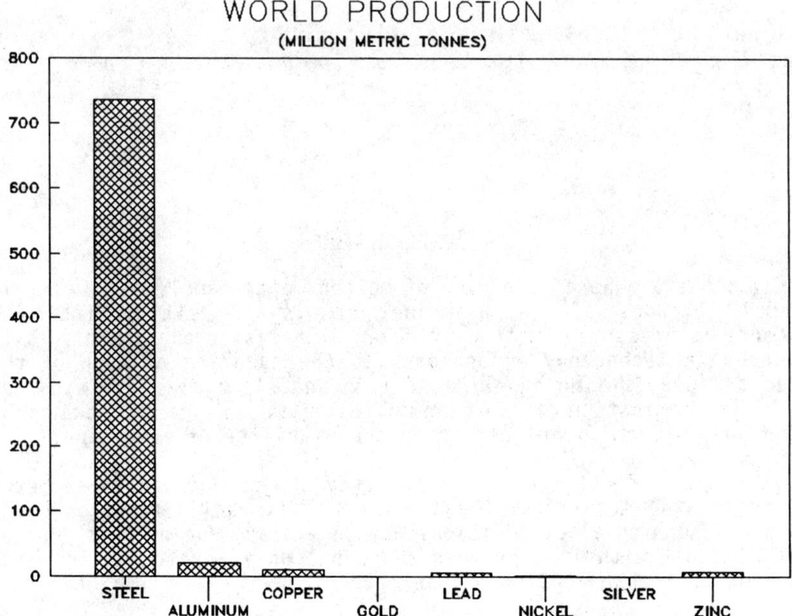

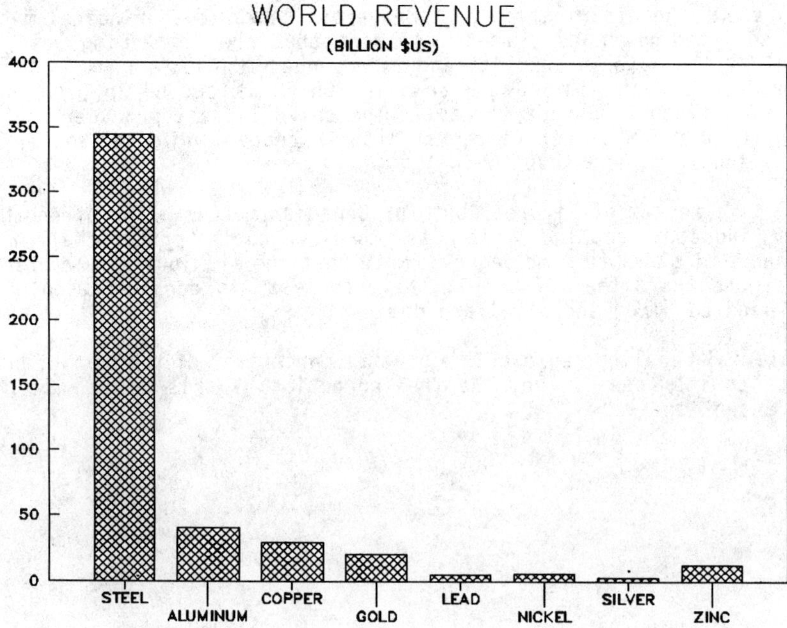

FIGURE 1: World Metals Production and Revenue

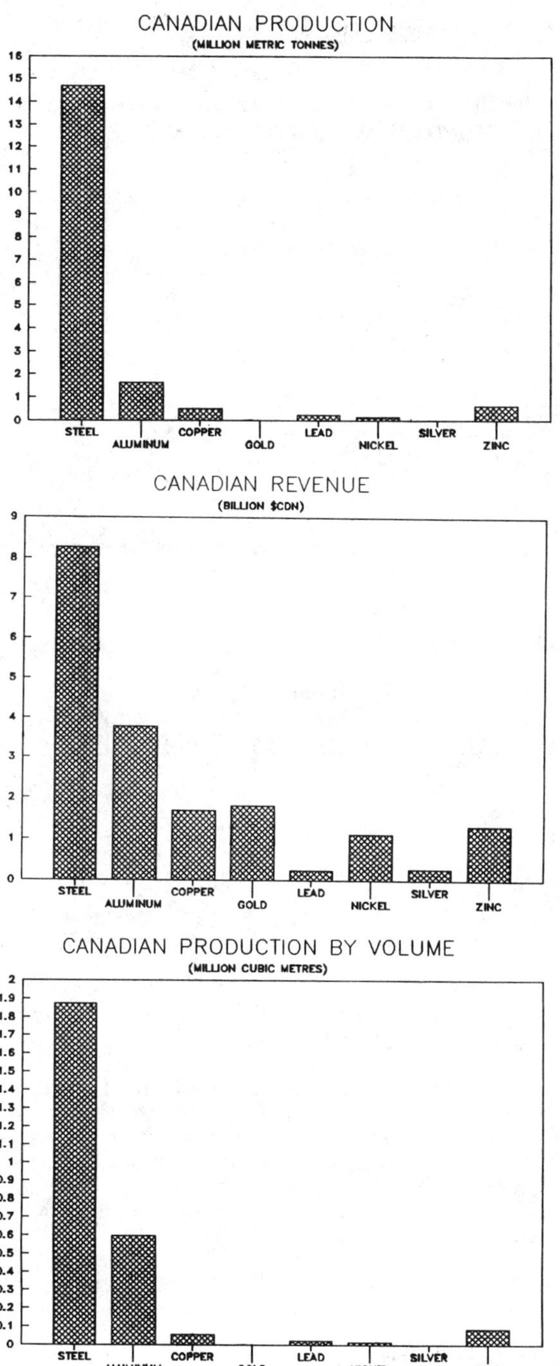

FIGURE 2: Canadian Metals Production, Revenue and Volume

Metal Characteristics

The following table demonstrates the distinctions between each metal which in part affect the method of molten metal transportation.

TABLE 1: Comparison of Metal Characteristics

	Melt Temp (°C)	Density (mt/m³)	Typical Ladle Capacity (metric tonnes)
Iron	1400	7.85	150 - 300
Steel	1650	7.85	50 - 200
Aluminum	650- 760	2.70	5 - 10
Copper	1150-1250	8.93	20 - 40
Nickel	1000-1600	8.90	20 - 40
Lead	400-1200	11.30	20 - 40
Zinc	400-1200	7.14	20 - 40

Table 2 shows an overall view of the various methods of transportation that are used in each industry.

TABLE 2: Comparison of Transportation Methods

	Ladles	Cranes	Runners	Rail Cars	Rubber-Tired Vehicles	Turrets	Pumping/ Piping	Syphoning
Iron	X	X	X	X	X			
Steel	X	X	X	X	X	X		
Alum.	X	X	X	X	X		X	X
Copper	X	X	X	X				X
Nickel	X	X	X	X				
Lead	X	X	X	X			X	X
Zinc	X	X	X	X			X	X

Iron and Steel

For every 100 tons of steel produced by the steelmaking furnace, 85 tons of iron is produced in the blast furnace, the remainder being made up of scrap. Given that exclusively scrap-based steelmaking represents about one third of crude steel production, iron handling effectively adds over half again as much tonnage handled before steel solidification. Thus it emerges that 88 percent of all molten metal handling in this country by tonnage, or 79 percent by volume, occurs in iron and steel plants.

Iron Handling

The batch processing of iron from the blast furnace to the BOF or electric furnace involves tonnages of up to 900 tonnes, hold times up to sixteen hours, travel distances up to two kilometres, and at least three separate types of conveyance: runners, torpedo cars, and ladles.

Torpedo Cars

The blast furnace typically produces 3500 tonnes per day. When this is tapped each of about six times per day, the molten iron runs down a runner where it is diverted into three or four torpedo cars (or "submarines"). These refractory lined, enclosed cars which travel by railroad to the steel mill are the mainstay of the iron shop. Although ladles are sometimes used directly the torpedo cars are easily the most common and most heat efficient mode of transportation in large plants.

Plant Layout

The physical distance that generally separates the blast furnace, which must have close access to ship or rail transport, from the steelmaking furnace, which must be close to the steel mill, initiated the reliance on rail transport. Although rubber-tired vehicles have the same flexibility and are currently being used in selected circumstances, only one ladle at a time can be carried. Furthermore, the long hold times detain the transporter (the rubber-tired vehicle) as well as the carrying vessel (the ladle). A rail line therefore continues to be the best option for iron transport.

Typically the torpedo car is emptied into a transfer ladle for desulphurization. The usage of low and mid alumina linings for the torpedo car will continue because the demands on the refractory remain low.

Weighing Devices

Only a few installations have attempted weighing the ladle in the transfer pit. One percent accuracy has been achieved, but damage is a serious concern and the maintenance requirements are unreasonable. If these problems were resolved, however, reliable tonnage weighing would provide significant improvements.

Environmental Considerations

Plant environmental standards will soon prohibit any visible emissions. The transfer of molten metal along the runners generated sizable quantities of iron dust and magnesium dust when most plants were built. Local hoods, ducting, and baghouse collection now control these fumes.

Steel Handling

Steel handling is characterized by high molten temperatures and large ladle capacities. These have accentuated concern for safety in overhead transportation and have focused more attention on refractory linings and heat dissipation. Also, because tap temperatures are higher the potential for energy savings is greater.

Molten Steel Transportation Methods

Molten iron and steel are almost exclusively carried by heavy-duty cranes and ladles. Newer approaches include ladle transfer cars, mobile rubber-tired vehicles, and ladle turrets.

In addition to providing rapid, reliable operation, the major advantage of cranes is their ability to move the ladle in all three axes. The disadvantage, which is now being addressed by other approaches, is the restriction of movement to the area of the crane runway. Crane handling cannot pass the ladle from one aisle to another and one crane cannot pass another on the same runway.

The increasing use of ground-level carriers and turrets has expanded the capabilities of movement through the shop or even off-site, but they are used to augment and not replace the cranes. Vertical lift is still necessary to reach the casting position.

Turrets are now used in two different instances, both of which make it possible to transfer the ladle from one aisle to another. In the first, the turret is mounted at floor level and has three stations: tapping the furnace, rotation into the ladle furnace, without the need for a crane, and a position from which the crane can pick up the ladle. With two independent arms, distinct operations can occur simultaneously. This compact design has encouraged single aisle layouts. In the second instance, a high level turret moves over the caster tundish and allows back-to-back casting by rotating a full ladle in as the empty ladle swings out.

Ladle transfer cars serve the same purpose as furnace turrets by their ability to transport ladles underneath the furnaces and to free the crane from having to transfer the ladle from one processing step to another. However, turrets and transfer cars are restricted in their motion as well, by the track or by the turret structure. Rubber-tired vehicles, on the other hand, are unrestricted in their movement except that they must be operated away from the furnace.

Rubber-tired vehicles are the newest method for molten steel handling, having been introduced only six or seven years ago. However, their use is growing as more casters are placed off-site. One example of the freedom that this provides in plant layout is when steel from multiple plants is cross-fed into a caster at just one of the shops. With side-shifting capabilities, a mobile carrier can even pick up a ladle placed by crane into a holding stand. Carriers have been designed to sustain loads of up to 450 tonnes including the ladle, its cover and contents.

Product Quality

The function of the ladle itself has changed from simply being a container for transporting molten metal into a metallurgical vessel of its own. By transferring the refining process to the ladle furnace the charging furnace becomes a pure melting device and both the throughput and the quality of steel are improved. In 1982, there were only six ladle furnaces in North America, in 1988 it was 30, and by 1992 it is expected that nearly all of the shops producing over 200 000 tons per year will have a ladle furnace. The ladle furnace is the final processing stage and quality can never improve after leaving it. Thereafter, material handling can only strive to diminish the reduction in product quality.

Another major trend in order to satisfy the increasing demand for clean-steel is to protect the hot metal from oxidation during transport. The use of ladle covers and shrouding between the ladle and the tundish and between the tundish and the caster mould prevent the metal from coming into contact with the air. Gas absorption and dross buildup are effectively eliminated. Ladle covers and shrouding also reduce heat loss significantly.

Ladle Design

The longer residence times and higher temperatures in the ladle resulting from ladle metallurgy have elevated the cyclic stress levels experienced by the ladle steel. This assertion is based on models developed to study the thermal and mechanical effects of these changes. Whereas refractories have changed significantly, ladle design has changed little. The next AISE specification for the design of ladles, however, is expected to incorporate more stringent guidelines. Vacuum degassing (VAD) techniques, which use suction to pipe liquid steel from the ladle, have put even greater loads on the ladle and its refractory.

To accommodate larger capacities while maintaining overall ladle height and trunnion centre distances, some plants have adopted obround or oval shaped ladles. While achieving the purpose, the structural integrity is affected and runouts have occurred. Nevertheless the need for tonnage expansion without high capital costs in plant conversion has encouraged this design.

The size of ladles has increased for two reasons. First of all, the general trend is towards fewer but more effective melting units along with facilities for ladle metallurgy. It is preferable to tap a larger heat into one ladle than it is to tap smaller heats into two ladles. Secondly, larger freeboards are now required due to the thicker slag layers that are formed in the refining process.

Heat Loss Reductions

Lids or covers for ladles are now more common because the tap temperature can then be lower, the time between taps can be longer, the tendency for skulling is decreased, and the heat loss through the slag is reduced.

Ladle Refractories

It is sufficient here to recognize that refractories have gone from bloating bricks to the more dimensionally stable dolomite bricks in order to accommodate the more severe operating conditions. Dolomitic bricks are longer lasting but have a lower coefficient of expansion. These stiffer refractories increase the demands placed on the ladle shell. Product quality is improved, however, because dolomitic bricks present no metallic contamination and contribute no oxygen to the steel.

A future possibility is the use of castable instead of bricks for the refractory lining. Since bricking practice varies from time to time and shop to shop, the modelling and analysis of the interaction between the ladle and the refractory should become easier with the single-unit castable.

Slidegates

At the teeming end of the production phase, stopper rods for ladle pouring have generally been superseded by slide gates. Acting as part of the process control loop, slidegates adjust the flow from the ladle and from the tundish. The rotary type has an advantage over the sliding type because the shape of the stream is always circular. In sliding types, where one hole overlaps another to restrict the flow, the nonlinear flow causes rapid wear. In both cases, the passage must be cleaned during every cycle and properly preheated, and the refractory must be replaced after every few heats. In general, though, the release of the metal through the slidegate is a cleaner, faster, better method.

Weighing Assemblies

Weighing devices, which were considered unreliable twenty years ago, are now used effectively on transfer cars, cranes and on casting turrets. Since loading of the carrier can therefore be better monitored, better safety conditions and less conservative designs are possible.

The predominant difficulty to overcome in weighing devices was keeping the load cells away from heat, impacts and horizontal forces. Load cells of the shear beam type, in which the stress of a deflecting beam is measured, have advantages in terms of ruggedness and ability to withstand horizontal forces. The new emphasis on weighing is motivated by the increasing use of computerized process control.

Economic Considerations

Cost savings can be made in three separate areas: capital costs, refractory costs and energy costs. The trend now, to save on capital costs and longterm planning, is to retrofit existing shops rather than build an entirely new infrastructure. Greater use of water cooling and shorter tapping times have reduced refractory consumption significantly. Meanwhile, additional advances in furnace design plus the use of ladle covers have reduced heat losses after melting. These can be easily reflected in lower energy costs. Waste heat has been utilized in some shops to further decrease these costs.

Non-Ferrous Metals

The transportation of molten metal in the non-ferrous industries embodies the same concepts as for iron and steel, but generally has the advantage of lower tonnages, lower operating temperatures, less tendency to skull and a strong trend towards continuous processing. The problems of heat loss, gas solubility, contamination by impurities, temperature control, plant layout, maintenance, safety, efficiency and environmental concerns remain.

Ladle Pouring versus Pumping

Aside from being time consuming and requiring a higher melt temperature because of heat dissipation, pouring from a ladle leads to dross and oxidation losses. By contrast, better quality metal is produced by pumping since the metal is brought from a submerged position and then transported in a closed environment. The transport time itself is also reduced as are the maintenance requirements. Molten aluminum, copper, zinc and lead can be piped over distances of up to 80 feet.

With respect to plant layout, pumping allows a centralized melting furnace to supply several pipeline systems. Furthermore, piping can be easily replaced.

Syphoned transfer is initiated by a small tilt of the ladle. It can be tilted upright again once the flow has started. Although the flow is difficult to terminate once it has begun, the syphoning technique provides a clean, quick transfer of the metal from the ladle.

Aluminum

Aluminum is the most versatile and castable of all the alloys and there are wide variations in volumes and methods of handling. Its low melting point and good fluidity allow for piped transfer by pumping and syphoning. Its low specific gravity, however, makes the separation of non-metallic impurities more difficult and the high solubility of hydrogen in aluminum makes close temperature control essential. Nevertheless, its much lower relative operating temperature makes cost effective high volume automated casting possible because supervision is not critical. Manual pouring is also viable due to the light weight of small and medium-sized aluminum castings.

The current use of an average of more than 120 pounds of aluminum wrought and cast products in each American-made car nearly triples the usage from ten years ago. This exceptional trend is certain to concentrate greater attention on all aspects of aluminum production in future years.

Copper, Nickel, Lead and Zinc

The traditional method during processing has been to transfer the metal by ladle between each stage. However, batch processing is inefficient, cranes are restricted to aisles, and fumes must be contained during the transfer. The modern trend in all cases has been to try to eliminate ladle transfer as much as possible. The processes, therefore, are now more closely coupled such that short launders connect each furnace or else several steps are incorporated into a single furnace such that no launders at all are required.

The greatest change in the handling of molten metal in the non-ferrous industries therefore has been to avoid it altogether by successively creating a more and more continuous process.

Conclusion

The present trends in hot metal handling are summarized as follows:

- Movement by pumping and syphoning is increasing in popularity;
- Dolomite bricks and castable linings are replacing traditional bricking practices;
- Movement by rubber-tired vehicles is becoming more common;
- Environmental concerns are requiring the increased use of hoods and covers; and,
- Electromagnetic stirring, pumping and molten stream shape control is becoming more widespread.

References for Figures

1. IISI, World Steel in Figures 1989 (1987 values).
2. World Bureau of Metal Statistics, World Metal Statistics Yearbook 1989 (tonnage based on primary and secondary refining).
3. Engineering & Mining Journal, 121st Annual Survey and Outlook, March 1990.

Hot metal handling in the aluminum industry

P.R. Whiteley and C. Simoneau
Alcan International Limited, Engineering Technology Group,
Montreal, Quebec, Canada

Abstract

Liquid aluminum produced in primary smelter reduction cells is handled by a variety of equipment before it is finally cast into solid shapes. The methods involve hot metal cranes to carry crucibles; siphons, launders, a variety of pumps, and road transportation.

Aspects which must be considered at all stages are safety, contamination, dross formation, temperature control, and cost. The article discusses Alcan's experience with the aforementioned equipment, and also covers the problems of climatic extremes which must be dealt with in Canada.

Introduction

In order to introduce the various techniques of metal handling, we will walk our way logically through the reduction and casting processes which take place in a typical Canadian smelter. The modes of metal transportation will be discussed at each step.

Since this is a paper delivered to a Conference of Metallurgists, we will endeavour to give it a metallurgical flavour by covering aspects of dross formation, metal purity and temperature control.

Liquid Metal Handling in the Reduction Step

Aluminum is produced from the oxide, alumina, by a reduction process performed in an electrolytic cell. The process is depicted diagrammatically in Figure 1.

Alumina is added, more or less continuously, into the hot liquid cryolite electrolyte in the cell (pot). The alumina dissociates into its constituent ions, and the positively-charged aluminum migrates to the negative cathode. The negative oxygen, on the other hand, moves to the positively-charged anode, whereupon it combines with the carbonaceous anode and leaves the cell as carbon dioxide.

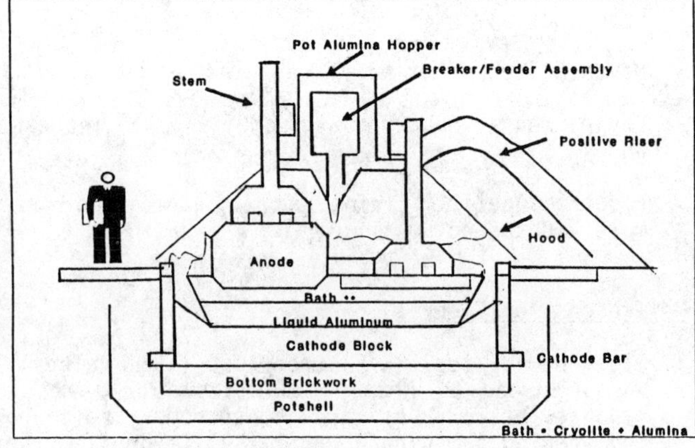

FIG. 1. Electrolytic Cell

While the production of metal in each cell is continuous, its removal is conducted on a batch-wise basis, typically once per day.

In modern cells, which range in size from about 150 kA to 300 kA, the quantity of metal produced each day ranges from one to two tonnes. Clearly, in order to produce respectable tonnages of metal, many cells are required. Figure 2 depicts a typical modern potroom which would accommodate about 100 cells. A modern smelter would commonly produce about 500 - 550 tonnes/day.

FIG. 2. Typical Potroom

The metal in the cell is so-called pure aluminum. It is derived from relatively-pure alumina, and so the only contaminants which arise are those which come in with the alumina or those which derive from the cell itself -- the electrolyte or the electrodes. Typical metal composition from the potroom would be:

Element	Percentage
Al	99.85
Fe	00.08
Na	trace
Ca	trace
Si	00.06

The metal is at 960°C when tapped.

Removal of metal from the cell is done by a vacuum crucible which incorporates a tapping spout. A multi-function electric overhead travelling crane carries the tapping crucible to the cell, lowers the spout through the electrolyte and down into the metal pad, and then the pressure inside the crucible is reduced by means of an ejector which is connected to a compressed air line adjacent to each cell. The rate of metal transfer by this means is approximately 1,350 kg/min and the corresponding compressed air consumption is 4.25 Nm^3/min.

Figure 3 depicts this tapping operation.

FIG. 3. Aluminum Being Syphoned from Cell

The cell being tapped is on line, and has a current of, say, 200,000 amps going through it. The problems of electrical isolation of the crucible from the crane, and from ground, are obviously areas of concern for electrical engineers.

The weight of metal tapped from each cell is indicated by a load cell on the crane hook and the weight is used for process and inventory control purposes.

The crane repeats the tapping operation until the crucible is full and the crucible is then taken by trailer to the casting plant. A variety of trailers are used, and the type depends largely on the method employed for transferring metal from the crucible into the holding furnaces.

A typical crucible transport trailer is depicted in Figure 4. The weight of the crucible full of metal is about 16 tonnes, and the trailer is designed for a speed of ten mph.

An unavoidable consequence of handling liquid metal from the potroom is that some of the electrolyte is occasionally tapped, and the aluminum oxidizes readily at these temperatures. The result of this is to create a gradual build up of frozen "bath" and dross on the refractory lining -- which, if allowed to go too far, would reduce the capacity of the crucible. A crucible cleaning station is therefore an attendant part of the liquid metal handling system.

FIG. 4. Potroom Crucible Transport Vehicle

Liquid Metal Handling in the Casting Plant

In the casting plant, the metal is transferred to holding furnaces where it is alloyed, fluxed, skimmed, settled, and the temperature set for casting. At the appropriate time, the furnaces are then tapped, usually by tilting, and the metal flows through slightly-declined launders through metal treatment devices to the casting machine.

The casting plant process is shown diagrammatically on Figure 5.

It is common practice within Alcan to, again, use the vacuum technique for transferring metal from the crucible to the holding furnace. This method results in a fairly gentle handling of the metal. It is done without a great deal of turbulence and without generation of new surface which creates instant dross and the consequent loss of valuable metal units.

In some plants, the crucible is simply upended, and the metal poured off the top rather like a teapot. The inversion of the crucible may be done by crane or by a trailer with suitable tilting apparatus.

Figures 6 and 7 show metal transfer devices from the crucible to the holding furnace.

The rate of metal transfer is typically one to two tonnes per minute, being faster with the tilting arrangement.

All of the current generation of casting plants within Alcan employ tilting/-holding furnaces in lieu of stationary furnaces. The tilting feature permits one to control metal flowrate to the casting machine with good precision through a closed-loop control system. It also permits one to stop metal flow quickly and reliably in the event of a problem. We use a variety of launder metal level sensors to control tilt rate, but the most common of these is a non-contact capacitance probe as shown in Figures 8 and 9. The signal from here is taken to a controller which adjusts the tilt rate to control the predetermined metal level. The actual metal flowrate is determined by the speed of the casting machine and the number and size of ingots cast simultaneously -- but is typically 700 kg/min.

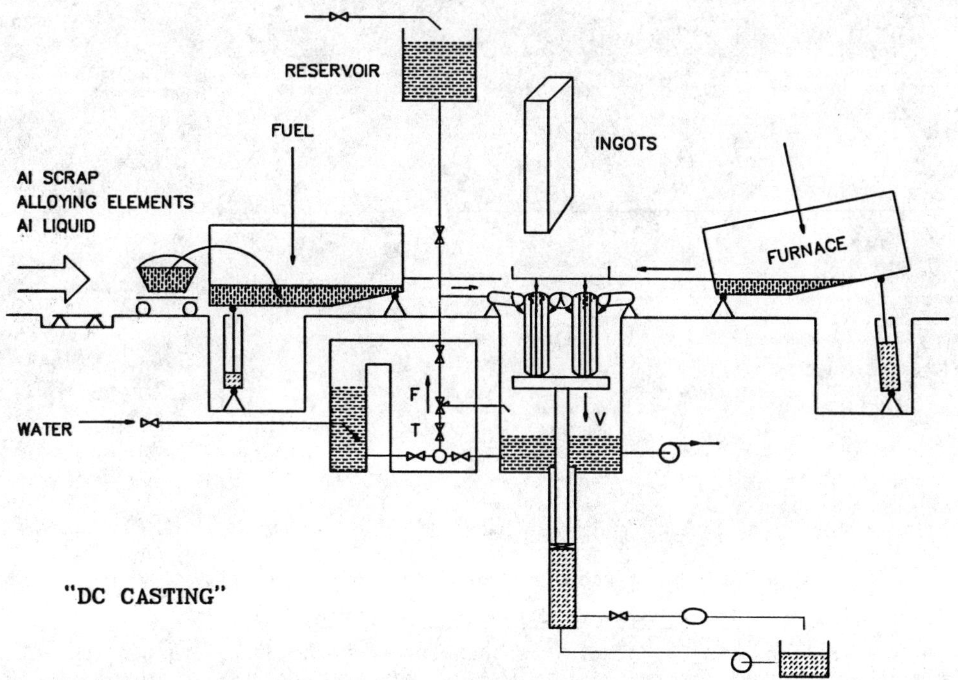

FIG. 5. DC Casting Plant Process

Open, U-shaped, insulated launders are used for metal transport from the furnace to the casting machine. The launder shape, insulation, slope and cross section are determined by a number of competing objectives:

- Minimum temperature loss along the launder length
- Maximum linear velocity without tearing the protective oxide film at the metal/air interface
- Minimum contamination
- Reliability and ease of cleaning

We select the cross section by calculating for an average metal velocity of about 10 metres/min. for the flowrate concerned.

As quality requirements and environmental considerations are leading us to in-line treatment of metal (as opposed to in-furnace) the launders linking the furnaces to the casting machine via these in-line systems are getting longer and longer (up to 20 m). In order to minimize the heat loss effect of the increased residence time inherent to longer troughs we have abandoned the rugged, but rather conductive (0.8 W/M•°K), low-cement castable liners in favour of pre-cast shapes of fused silica foam or of Insural (a proprietary mixture of Wollastonite and silica cement).

While fused silica foam has a low conductivity (.15 W/M•°K) and a good resistance to the chemical attack of molten aluminum, it has the drawback of having a low tensile strength and very low coefficient of thermal expansion. Because of these characteristics FFC shapes must be installed in water-cooled shells to prevent differential expansion from destroying the costly lining.

FIG. 6. Metal Transfer by Crucible Tipping on a Crane

FIG. 7. Metal Transfer by Crucible Tipping on a Vehicle

However, water-cooled shells are costly (approximately \$2,500/m), and require frequent maintenance. In recent installations, Alcan has used 50 mm thick Insural pre-cast shapes which are laid on a 25 mm thick bed of ceramic fibre wool into non-cooled steel shells. Insural has a conductivity of .2 W/M•°K and a coefficient of thermal expansion ten times greater than FFC. The soft and insulating ceramic fibre wool first allows the reduction of combined conductivity of the two-layer liner to .15 W/M•°K, and secondly permits differential expansion of the air-cooled steel shell to occur without damage to the rigid inner liner; obviously, expansion joints are provided to prevent distortion of the metal shell.

At the end of a cast, there is always a paper thin layer of aluminum which solidifies on the walls of the launder; this must be removed to prevent alloy-to-alloy contamination. The walls are then daubed with a suspension of inert material which performs the double function of protecting the insural liner from the corrosive action of molten aluminum and of acting as a parting agent.

FIG. 8. Capacitance Probe

The metal temperature in the cells in the potroom is 960°C, and after tapping and transportation it is usually of the order of 850°C by the time the metal is transferred to the casting plant. The casting temperature is about 700°C, and so the metal temperature actually falls in the holding furnaces. There is, however, an unavoidable amount of run around scrap in the casthouse, and the remelting of this solid metal and radiation and convection losses from the holding furnaces usually result in there being a net energy input to the furnaces, notwithstanding the initial superheat of the molten metal charge.

FIG. 9. Probe in Position

Other Methods of Molten Metal Transportation

In most cases we design plants so as to be able to use gravity for metal flow. In some older plants, and for the purposes of metal recirculation however, metal pumps are used to a limited extent.

FIG. 10. Metaullic Metal Pump

The first of these is a centrifugal pump, of which the Metaullics brand is typical (see Figure 10). Pumping liquid metal is obviously an arduous function, and aluminum is particularly aggressive and reactive with most metallic elements. Accordingly, the metal contact components are graphite, but even this material is subject to fairly rapid erosion. Plants employing this type of unit usually have one or two standby units and require a consistent maintenance effort. These pumps are frequently employed to circulate metal within a melting furnace for the purposes of improving melt rate through forced convection. They may also be used with refractory-lined pipelines for metal transportation although Alcan does not use pumps for this purpose on a production scale.

A different pump is the electromagnetic type shown on Figure 11. The advantage of this is that there are no moving parts, nothing except refractory in contact with the metal, and maintenance costs are reported to be low. Their use as a pump in the conventional sense of the word is clearly limited, although they are commonly applied as a furnace metal circulation device.

The capacity of both of the aforementioned devices is in the order of 1,000 kg/min.

A third type of "pump" is the alternating-pressure pump which is, again, used strictly for metal circulation within a furnace, not for bulk metal transportation. This pump has the advantage of much higher instantaneous flowrates than the previous types, resulting in very rapid homogeneity of alloy elements. A Japanese version of this type of device is depicted on Figure 12. Alcan has developed its own style of patented pump, operating on similar principles, and it is used successfully on a number of furnaces in Canada.

Highway Transportation

Alcan has a very extensive series of operations in the Saguenay region of Quebec, involving smelters and casting plants. In order to balance the needs for metal, we transport liquid metal over the public road system with specially-designed vehicles.

Between 75,000 and 100,000 tonnes of liquid metal are transported on public roads every year. Transport distances vary from 30 to 100 kilometres and allowable speed on the routes followed varies from 30 to 90 km/hr.

Figure 13 shows a road transport crucible firmly secured to the depressed bed of an otherwise standard road-certified tractor-trailer arrangement.

The principal dimensions of the container are:

- Outside diameter 2.6 m
- Overall height 2.6 m
- Height to lid line 2.1 m
- Internal diameter 2.25 m

The crucible itself weighs 12 tonnes and can hold up to 14 tonnes of molten aluminum, that is, the contents of two potroom crucibles. Its lining has been designed to minimize the rate of heat loss as well as to resist the erosive action of the free falling metal as it is being transferred from the tapping crucible. The selection of the lining materials was also influenced by the need to keep the weight as low as safely possible (axle load limits are greatly reduced during the spring thaw season in Quebec).

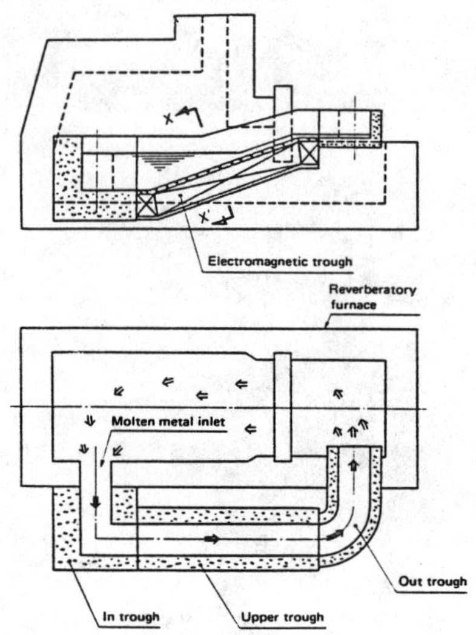

FIG. 11. Electromagnetic Pump

Heat losses on the road, expressed in terms of metal temperature drop, are of the order of 8 to 10°C/hr. We find that the extreme winter temperatures in which we operate (-40°C, or snowstorms) have little effect on the rate of heat loss; paradoxically, a severe rain storm will increase the heat losses to 18 - 20°C/hr. The greatest sources of heat losses in this operation are the transferring of metal from potroom crucible to transport crucible and re-heating of the road crucible lining. On average, the potroom metal will arrive at its destination at 780 to 810°C. The variation derives less from the distance of transport than from the time the crucible is left idly cooling, waiting for metal. Experience demonstrates that a transport crucible can safely be left idle for a period of up to eight hours; beyond that, it must be reheated.

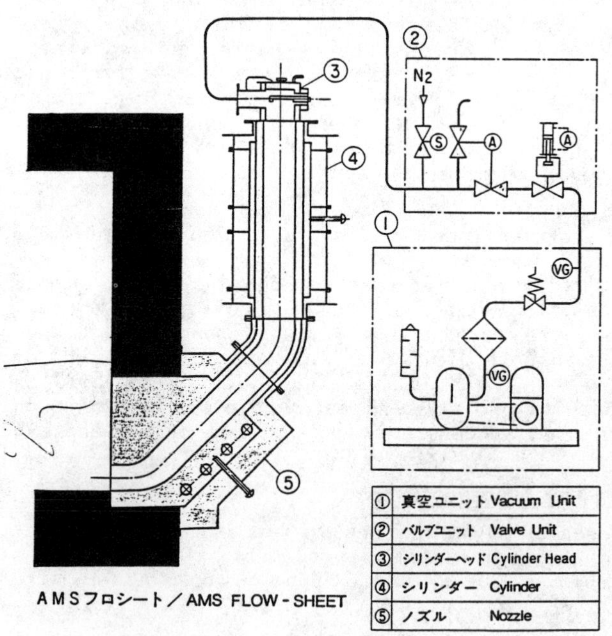

FIG. 12. Push/Pull Melt Stirrer

This method of transportation has proven to be safe and reliable. In eight years at 14 to 18 trips per day, we have yet to report a single accident. Figure 13 illustrates the vehicle and crucible assembly. Measures are taken to first secure the crucible on the truck bed, then to secure the tap hole pins, lid and filling hatch cover to the crucible.

FIG. 13. Road Transport Vehicle & Crucible

Everywhere in the design of this transport system, very conservative safety factors have been used.

Safety

Last, but certainly not least, is the general aspect of safety.

Most metals exhibit an exothermic reaction when the metal is converted to its oxide. This is especially so with aluminum, and under certain conditions, the oxidation can result in a powerful explosion. Certain substances, such as water, iron oxide, and nitrates, can, when combined with liquid aluminum, lead to explosions. Accordingly, there are very specific guidelines adopted in the aluminum industry to guard against mishaps. The best known document in this regard is the "Guidelines of Handling of Molten Aluminum", published by the Aluminum Association in the United States.

This volume covers all of the specific procedures and practices to be adopted in melting, transportation and casting of liquid metal. The most common cause of problems is the entrapment of water under aluminum. Even quite small quantities of moisture are sufficient to cause "pops" which could result in burns to operators and damage to equipment. Particular care must be taken to guard against wet scrap being charged into furnaces -- especially moisture in shrinkage cavities of remelt ingot. Furnace tools and newly-repaired refractory work must be properly preheated and dried before coming into contact with metal.

In addition to the foregoing one has to address the usual metallurgical industry problems of protecting people and equipment from liquid metal and hot surfaces. Aluminum is potentially difficult in this regard, because, unlike molten steel or iron, it maintains a cool silvery (not red) appearance when liquid.

A reassessment of the role and design of hot metal launders

M.F. Shaw
Bechtel Corporation, Mining & Metals, San Francisco, California, U.S.A.

Introduction

The transfer of hot metal has been an integral part of metallurgy since the beginning of the Bronze Age. The first recognizable transfer might have been created by a paleo-pyrometallurgist scooping a trough in the dirt so that his crudely smelted copper was diverted away from the smelting furnace. In this manner, the product could be obtained without destroying the furnace. The trough technology probably had roots in the agricultural heritage of this early man. This trough would have evolved quickly to a trough with a pit at the end to produce a remeltable button or shape more suitable for subsequent processing such as hammering, drawing or the like. As the technology for casting advanced, the pit probably evolved into a pot and the pot eventually became a less fragile ladle. The trough connecting the furnace with the ladle remained an integral part of pyrometallurgy. This trough is generally called a launder and, in many instances, has changed little over the centuries.

This paper will give an overview of the types of launders used in representative nonferrous and ferrous industries and will list some developments which have replaced the launder and will discuss possible future launder innovations or alternatives.

Discussion

A launder as a means of transferring hot metals from one vessel to another has the advantage of simplicity and low cost. It also has some inherent disadvantages:

- limited useful lifetime
- usually must be heated before and/or during transfers
- material being transferred is exposed to air and possible detrimental oxidation
- possibly toxic, corrosive or valuable volatile components of the molten metal can be emitted to the working environment

- As a gravity device, it constrains plant layout and requires elevation differences between operations.

Unfortunately, as the molten metal transfer temperature increases the possible alternatives to the use of a launder decrease. The development of molten metal transfer schemes, for metals having lower transfer temperatures, has been confined to pumping or siphoning. The advances made in transfer schemes for higher transfer temperature metals have been sophisticated materials for construction of the launder.

A review of current usage and substitutions for launders in various metal industries follows. The copper industry has retained launders more than any other because of the relatively low reactivity in air of the materials transferred in a molten condition. The temperatures of these materials is high enough to make pumping difficult because of materials of construction limitations. The steel industry has limited launder transfers due to the extreme temperature of the molten material and the deleterious effect on most materials of construction. The predominant means of transferring hot materials in steel making is direct ladle transfers. The lower melting metals industries such as lead, tin and zinc now pump hot metals which is superior to launder transfers. The aluminum industry uses some pumping and avoids open launders because of the high reactivity in air of molten aluminum.

Copper

Because of the moderate reactivity in air and high-transfer temperatures (1100-1300°C) of copper, the copper industry has the greatest diversity of launder applications. Molten copper is too hot to pump with commercially available materials of construction. In the copper industry the present trend is away from traditional batch processes such as Peirce-Smith converting to continuous smelting and converting operations. These newer processes require increased reliance on launders as a means of transferring hot metals and slags from vessel to vessel.

The materials used depends primarily on the reactivity of the materials transferred and the longevity of the *p896Xlmaterial in this particular application. Applications are often specific to the types of materials locally available. The first molten metal material encountered in conventional coppermaking is called *matte*, which is a copper-iron sulfide intermediate product. Matte is usually transferred from the smelting furnace into a ladle for transfer to a batch converting operation. State-of-the-art copper smelting technologies feature a continuous transfer of material from a continuous smelting furnace to a continuous converting furnace. One advantage of these continuous processes is that expensive overhead cranes, used for the ladle transfer from vessel to vessel, are eliminated. Matte has an property, which is shared with aluminum, in that it can create violent steam explosions on contact with water. This precludes water-cooled launders which would reduce the erosion of the materials of construction by the molten matte if they could be used. Matte also dissolves steel and iron. The materials in matte launders are usually a steel trough with a copper, steel brick, refractory or graphite liner. Solid copper blocks are also used.

Copper smelting slag, although hotter than matte, is more easily handled. Water-cooled copper, stainless steel or graphite launders are used for the transfer of copper slag. The graphite launders are common in Japan where the eroded graphite matte launder-liners are reused as slag-launder liners.

Water-cooled copper block launders have been traditionally produced by pouring molten copper over a steel pipe coil in a mold that is the shape of the launder. Sometimes a steel shell is used instead of the mold and the steel is stripped away at the launder-slag interface once the copper has solidified.

The state-of-the-art for water cooled copper launders is the use of continuously cast copper blocks with drilled and plugged water passages. This allows superior heat transfer, eliminates the high rejection rate during launder manufacture and eliminates costs of casting copper onto the steel pipe coil launder.

The final molten copper product is generally laundered with refractory-lined steel troughs since this steel maintains the temperature of the molten material better than do some of the water-cooled launders. Heating is often required on open launders. A rate of 0.7 MJ per meter of launder can be used and if the burners are set fuel rich the copper is less likely to oxidize during the transfer.

Launders in the copper industry have generally been kept short, less than 6 meters long, because of the drop necessary to keep molten material flowing without excessively superheating the molten metal. The elevation drop with distance (5%) can equate to a substantial elevation difference between vessels that can be very expensive.

Although the trend in the lower melting metals has been toward pumps and siphons, the copper industry has not adopted either of these technologies. A centrifugal pump of the proper materials of construction could have a great benefit. As environmental requirements become stricter there is a need to collect all fumes from the various hot metal transfers in coppermaking. These environmental criteria can best be met by containing the molten metals in refractory lined pipes and pumping or siphoning from one location to another. The same development of suitable materials for launders could aid in the manufacture of a vertical centrifugal pump for copper. The impregnated graphite used in zinc and aluminum pumping has also worked in static applications in coppermaking so a graphite pump for copper is a possibility.

Lead

Launders generally find application in the primary and secondary lead industry only in the short run from the smelting or melting furnace into the first drossing pot. In this case, cast iron has been the standard material of construction used for the launder. Covers are used and the launder is not heated.

The primary means of transferring lead and lead alloys from vessel to vessel or mold is with pumps. Vertical centrifugal pumps are used with an impeller that is cantilevered to keep the bearings out of the hot metal. Temperatures in the range of 370 to 510 °C are common. Mild steel or black iron has been the standard materials of construction for both pumps and transfer piping because the lead and lead alloys generally encountered are not corrosive to these materials. Stainless steel has been tried, but the nickel leaches into the lead and renders the material useless. Use of high head pumps keep velocities high. The transfer pipe is often preheated prior to effecting a transfer but is not heated during a transfer. Transfer piping runs of 6 meters are common and 12 meters is not unusual.

The segregation of alloying elements during pumping has not been a problem. The return pumping of molten material is occasionally used to melt scrap material through forced circulation.

Zinc

Zinc and zinc alloys are another example of a metal with a lower transfer temperature. The use of a launder for zinc has several disadvantages in that the metal fumes and oxidizes during an open air transfer. The dross formed from air oxidation requires reprocessing for metal recovery. Differential oxidation will changes alloy compositions making metallurgical quality control more difficult. Pumps have found utility in the zinc and zinc alloy industry where they are used between breakdown vessels, refining and holding furnaces, casting units and ladles. Pumps have also been used to empty galvanizing kettles. The pumps used are generally similar to the type of pump used in the lead industry in that they are vertical centrifugal units driven by air-motors. The pumps are more sophisticated than lead pumps because they can handle material that has a higher temperature and more corrosive nature. Since the pump is submerged, it will not transfer dross, skim or fluxes. Pumping can also transfer molten materials more quickly than pouring can.

The molten metal transfer temperature of zinc and zinc alloys (550 °C) is somewhat higher than lead. Additionally zinc and its alloys are corrosive to some materials of construction. These factors have lead to the development of two types of materials for this application: graphite and austenitic stainless steel. Modifiers are added to the graphite to prevent oxidation at the pump operating temperatures. The graphite pump is suitable for use with pure zinc or zinc alloys while the stainless steel pump cannot be used with pure refined zinc or with galvalume alloys. Pump capital cost becomes the criteria when either of these materials of construction can be used. Other materials of construction have been tried, and only a molybdenum-tungsten alloy has been successful. This alloy is very brittle which limits its application.

Zinc is commonly transferred by a pump to a subsequent vessel or furnace in open refractory launders of up to 15 meters in length. The heat lost in this launder is sometimes replaced by radiant heating elements in the covers of over the launder. The pumps are used to allow the furnaces and vessels to be at the same floor level. The transfer launders are subject to the previously mentioned oxidation and fuming problems. The trend in this application is toward use of refractory-lined pipes where distance and layout are not a factor.

Aluminum

The reactivity of molten aluminum in air, has generally precluded the use of open launders. When transferring molten aluminum from a smelting pot to the casting operation, a ladle is used that has a spout inserted into the smelting pot. A vacuum is applied to the ladle and the molten aluminum is sucked out of the smelting pot into the ladle. The molten aluminum is then carried to the casting furnace which may be some distance away. The top of the ladle is removed and the material is poured into the holding furnace. A newer scheme is to pressurize the transfer ladle with air and force the aluminum down the spout. This eliminates the need to reseal the top of the ladle and eliminate the oxidation which occurs while pouring.

A short cast iron launder connects the casting machine with the holding furnace.

Pumps are also finding application in the aluminum industry in remelting and refining operations for forced circulation and degassing. The construction of these centrifugal air motor-driven vertical pumps is similar to those used in the zinc industry. A graphite which has been stabilized for the operating temperatures is used as are ceramic coatings and linings. Refractory-lined pipelines have been used on these pumping applications.

Steel

Steel, the metal capable of reaching the highest temperatures (1500°C) considered in this paper, uses batch transfer of molten material by refractory lined ladles, sometimes by pouring hot metal out the bottom of the ladle. The only general application of launders in the iron and steel industry is the transfer of iron and slag from a blast furnace to torpedo cars or slag granulators. In this case a substantial development program has been undertaken to discover those materials which will survive the conditions imposed by hot molten material streams striking the refractory surfaces.

To facilitate maintenance and repair of the launders, a design has evolved that consists of a steel refractory-lined trough within a trench which allows the entire assembly to be removed and replaced with minimal furnace downtime. The trench is wider than the launder so air cooling allows a longer life.

Conclusions

The launder has been around a long time and promises to remain in limited applications in the future. Environmental concerns to capture all fugitive gases evolved from a process limit the potential for launders. The pumping of hot metals is becoming more common as advances are made in materials of construction.

Bibliography

D. V. Neff, "Pumping and pipeline transfer of molten zinc alloys," TMS Technical Paper No. A89-10.

R. K. Shoemaker, "Casthouse Refractory Practices Armco's Amanda Blast Furnace at the Ashland works." ARMCO steel 1988.

Thermomechanical aspects of lining behaviour as related to design, installation, and use in ladles

C.A. Schacht
Schacht Consulting Services, Pittsburgh, Pennsylvania, 15235, U.S.A.

Introduction

Relining, lining maintenance, and ladle repair can represent a significant cost in steelmaking. Therefore, the objective of all steelmaking shops is to minimize this cost and to maximize the lining life. However, the thermomechanical behavior of the ladle lining is highly complicated and sometimes misunderstood. The ladle design also has a significant influence on lining life. Other factors which have a direct influence on lining life are the shop lining installation practices and manner in which the ladle is used. The objective of this paper is to provide an understanding of the impact of these factors as related to improving the integrity of the lining and increasing lining life.

Fundamentals of Steelmaking Ladle Lining Expansion Forces

Numerous studies have been conducted by the author on steelmaking ladles with molten metal capacities ranging from 110 to 350 tons. The steelmaking ladle typically has a sidewall plate with a slight taper and for analytical purposes can be classified as a cylindrical vessel.

A typical ladle is described in Fig. 1. The ladle sidewall lining usually consists of a working lining and a safety lining. Although the type of refractory used in the working lining varies based on the type of steel produced and other metallurgical requirements, the refractories typically used are: 70 percent alumina brick; direct bond mag-chrome brick; resin bond dolomite brick; and direct bond dolomite brick. Fireclay brick, high alumina brick, or a similar strength brick is used as the safety lining.

Normally, the ladle lining is exposed to a through-thickness temperature gradient and restrained by the ladle sidewall plate, is described in Fig. 2. The hot face side of the refractory lining develops a compressive thermal force and the ladle sidewall plate develops an equilibrating tensile force. The radial joints open on the cold side of the working lining as the lining joints have little or no tensile strength. A radial compressive load (P) develops between the working lining and the vessel shell due to the thermal restraint imposed on the lining by the shell. The radial compressive load is transferred from the working lining through the safety lining onto the ladle sidewall. Therefore, the safety lining must have the strength to accommodate the radial compressive load (P).

Refractory Stress/Strain Behavior

An important requirement in evaluating thermal expansion stresses in ladle linings is to have defined the compressive stress/strain behavior of the working lining refractories. This data is necessary in order to accurately determine the magnitude of the thermal compressive stresses in the working lining and the equilibrating tensile stresses in the ladle sidewall plate.

Since research on thermomechanical behavior of refractory materials is very limited, the impact of the multi-dimensional stress states and cyclic loadings is not fully understood. Studies on similar materials, such as structural concretes, are more extensive and provide some insight into the thermomechanical behavior of refractory materials. However, refractory materials unlike structural concretes are exposed to temperatures in excess of 1700^{0}C.

Since refractories have high compressive strength and low tensile strength, compressive stress-strain behavior of refractories forms the major portion of the total strain energy developed in the refractory lining. Therefore, it is important that the compressive stress-strain behavior of the refractory material used in lining the cylindrical vessel is understood.

Static compressive stress-strain data for various refractory materials are described in Figs. 3 through 6. As shown, the high alumina (98 percent) castable and the direct bond mag-chrome brick are considerably stiffer than the high alumina (70 percent) brick and the resin bond dolomite brick. Mortar joints tend to reduce the stiffness of a refractory lining system as shown in Fig. 7.

Bilinear stress-strain curves are used in computer models to simulate the nonlinear stress-strain curves of the refractory material. As shown in Fig. 8, a significant part of the bilinear stress-strain fit matches the actual nonlinear stress-strain curve for the anticipated strain range imposed on the refractory lining material.

<u>Details of Lining Expansion Forces</u>

The expansion of the ladle wall working lining relative to the shell plays an important role in preventing molten metal penetration. The expansion of the working lining relative to the shell creates a mechanical (1) seal on the working lining hot face (see Fig. 2). Under complete restraint, the potential temperature dependent expansion force, F(T), that is developed by the refractory working lining material is defined as:

$$F(T) = E(T)\alpha\Delta T \qquad (1)$$

where: E(T) is the temperature dependent elastic modulus;   is the coefficient of thermal expansion; and ΔT represents the change in temperature from ambient conditions.

As described in Eq. 1, the elastic modulus and the coefficient of expansions are the two material properties of the refractory that determine the magnitude of the expansion force. Therefore, for the purpose of ranking the potential expansion force of a given refractory (assuming the same ΔT for all refractories), Eq. 1 can be written in the form:

$$F(T) = E(T)\alpha \qquad (2)$$

A comparison is made of the elastic modulus of the six types of refractory brick of interest in Fig. 9 (2,3,4). As shown, the direct bond-chrome has the greatest elastic stiffness for operating temperatures above 815^{0} (1500^{0}F), while the soft 70 percent alumina and the resin bond mag-chrome have the least elastic stiffness.

The calculated values (defined by Eq. 2) of the potential expansion force of the six types of refractory brick are plotted in Fig. 10. Up to an operating temperature of approximately 925^{0}C (1700^{0}F) the expansion force capability of the direct bond dolomite brick is the highest; whereas, above 925^{0}C, the direct bond mag-chrome brick slightly exceeds the potential expansion force capability of the direct bond dolomite brick.

The expansion force data shown in Fig. 10 implies that zoning with different types of refractory brick may result in considerable differential expansion between the zoning brick and the working lining brick. Of concern in the design of zoned linings is the mismatch between expansion force of the adjacent brick in the region of the zoning. For example, the use of a direct bond dolomite zone adjacent to a soft 70 percent alumina working lining can result in joint openings in the working lining with molten metal penetration in the region surrounding the dolomite zone.

It should be noted that the most severe mismatch of potential expansion force of the refractory occurs in the range of temperatures from a few hundred degrees centigrade to about 1200^{0}C.

<u>Wall Lining Expansion Allowance</u>

As previously discussed, the magnitude of the expansion force varies considerably based on the type of refractory used in the working lining. Also, all of the refractory brick, except the direct bond mag-chrome, have considerable loss in stiffness in the portion of the lining with temperatures above 1800°F. For the direct bond mag-chrome brick, the loss in stiffness does not occur until the lining temperatures exceed 2300°F.

The expansion difference between the lining and the shell can be interpreted as an interference fit. For a given type of refractory working lining, the interference may differ based on the thickness of the safety lining used between the working lining and the shell, as well as other factors which influence the lining and shell temperature profile.

In order to evaluate the need for a tight lining installation, it is worthwhile to estimate the interference for the various types of refractory addressed in this discussion. For the example used here, it is assumed that the ladle wall lining geometry and operating temperatures are identical for each type of refractory considered.

The interference for the six types of refractory working linings considered is shown in Table I. As shown, the 70 percent alumina working lining has the least interference, slightly greater than 1/8 inch, and requires more care during installation than the other refractory linings. This implies that a tight installation is required for the alumina lining in order to avoid molten metal penetration into the brick joints. Also, it implies that compressible insulation board or blanket material could not be successfully used with a 70 percent alumina working lining as the interference would be lost due to the compressibility of the insulating material resulting in a loose lining.

TABLE I

Interference of Working Lining

Type of Working Lining,	Interference, inch
70%, Al_2O_3 (Both Stiff and Soft)	0.145
Direct Bond Mag-Chrome	0.397
Resin Bond Mag-Carbon	0.36
Resin Bond Dolomite	0.487
Direct Bond Dolomite	1.401

The remaining types of refractory working linings described in Table I have considerably greater interference than that 70 percent alumina lining, implying that less care is required during installation without sacrificing loss of the mechanical seal in preventing molten penetration.

Heating Cold Ladles

Ladles are typically preheated by natural gas ladle heaters with hot gas temperatures reaching maximum values of about 1100°C (2000°F). Preheat of the cold ladle can be done in a "ramped" manner of about 95°C/hour or in a "rapid" manner of about 1100°C/hour. After preheating the ladle to near maximum steady-state temperature, it is placed into service carrying molten metal in the steelmaking shop.

The results of a strain gage measurement (5) on the teeming ladle lined with a 70 percent alumina brick working lining are summarized in Fig. 11. A rapid preheat was used which resulted in vessel shell expansion stress of about 16 ksi. As shown, the analytical expansion stress results agree with the field test results.

When stronger types of refractory are used as a working lining, such as the direct bond mag-chrome, then the resulting shell stresses can exceed the yield strength of the sidewall plate.

Ladle Shell Flexing

Another type of stress imposed on the lining are the mechanical stresses resulting from ladle shell flexing (6). In the case of a ladle, shell flexing (see Fig. 12) occurs when a ladle containing hot metal is held at the trunnion pins by the crane hooks. Cyclic mechanical stress results from multiple ladle lifts and set downs which are necessary to satisfy the needs of ladle metallurgy and caster operation. In ladles, the new clean steel metallurgical practices often require considerably more ladle lifts and set downs than experienced in the past. As a result, the working lining is exposed to considerably more flexing.

The ladle stiffener bands provide the necessary strength to support the ladle by the trunnion pins. However, it will be shown that the size of the stiffener band (and the resulting stiffness of the shell wall) varies greatly from plant to plant. This implies that the cyclic mechanical stress in the lining will vary from plant to plant.

The results of a limited investigation of several different steelmaking shops with molten steel breakouts in the ladle wall are summarized in Fig. 13. The ladles were lined with 70 percent alumina brick. Ladles with more flexible stiffener bands experienced the greatest frequency of breakout.

Summary

In summary, the following primary areas of ladle/lining system should be addressed to reduce lining maintenance and to improve the integrity of the ladle/lining system. They are:

- For the type of refractory working lining selected, use an appropriate thickness of compressible insulating material to avoid over stressing the working lining and ladle shell.
- For high alumina working linings, compressible insulating material should not be used.
- For high alumina working linings, tight lining installation is an absolute necessity.
- Avoid using low strength rigid insulating brick when using the high strength refractory working lining.
- Avoid rapid heating of a cold ladle.
- Make certain the ladle has adequately sized stiffener bands. The top stiffener band is often underdesigned with regard to flexing during ladle lifting with full molten metal.
- Minimize ladle lifts while the ladle is filled with molten metal.
- When zoning is used, select compatible types of refractory materials that minimize differences in thermal expansion behavior.

References

1. C.A. Schacht, "Update on Factors Contributing to Molten Metal Breakout in Iron and Steelmaking Ladles," Abstract, Iron and Steel Engineer, Vol. 64, No. 9, AISE, Pittsburgh, PA, 1987, pg. 82.

2. M.A. Hartsock,"Stress/Strain Data on Ladle Brick," Technical Memorandum, Kaiser Aluminum and Chemical Co., Pleasanton, CA, July 6, 1983.

3. R.C. Kaldon, "Data on Harbison Walker Nuline R20SA," Harbison Walker Refractories Co., Pittsburgh, PA, May 21, 1985.

4. C.A. Schacht, Report on Investigative Study on the Thermomechanical Behavior of Resin-Bonded Dolomitic Brick Working Lining, AISE Subcommittee No. 9, AISE, Pittsburgh, PA, March 6, 1987.

5. C.A. Schacht and E.V. Abarotin, "Structural Behavior of Teeming Ladles Lined with High-Alumina Refractories," Iron and Steel Engineer, Vol. 61, No. 8, AISE, Pittsburgh, PA, 1984, pp. 33-40.

6. C.A. Schacht, "Lining Life of Ladles as Related to Shell Flexing and Lining Properties," Iron and Steel Engineer, Vol. 64, No. 6, AISE, Pittsburgh, PA, 1987, pp. 48-52.

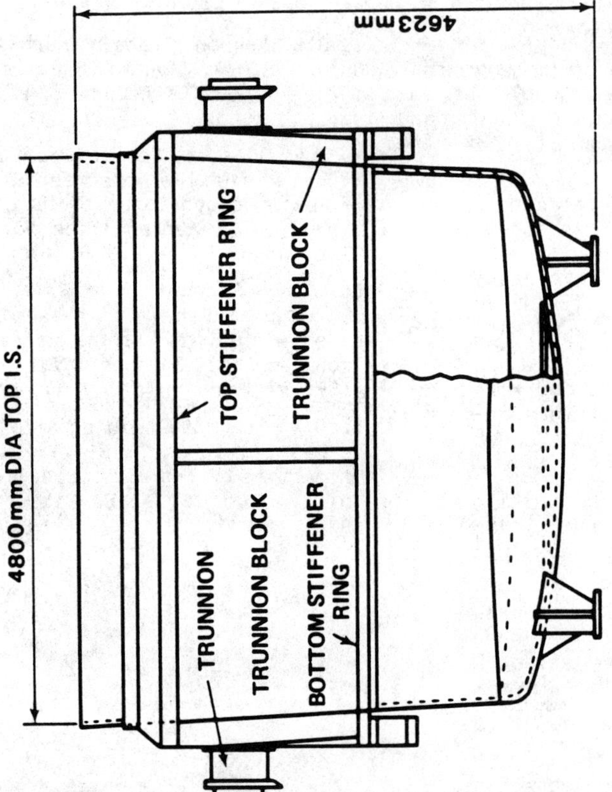

FIG. 1. 315 to 350 Ton Teeming Ladle

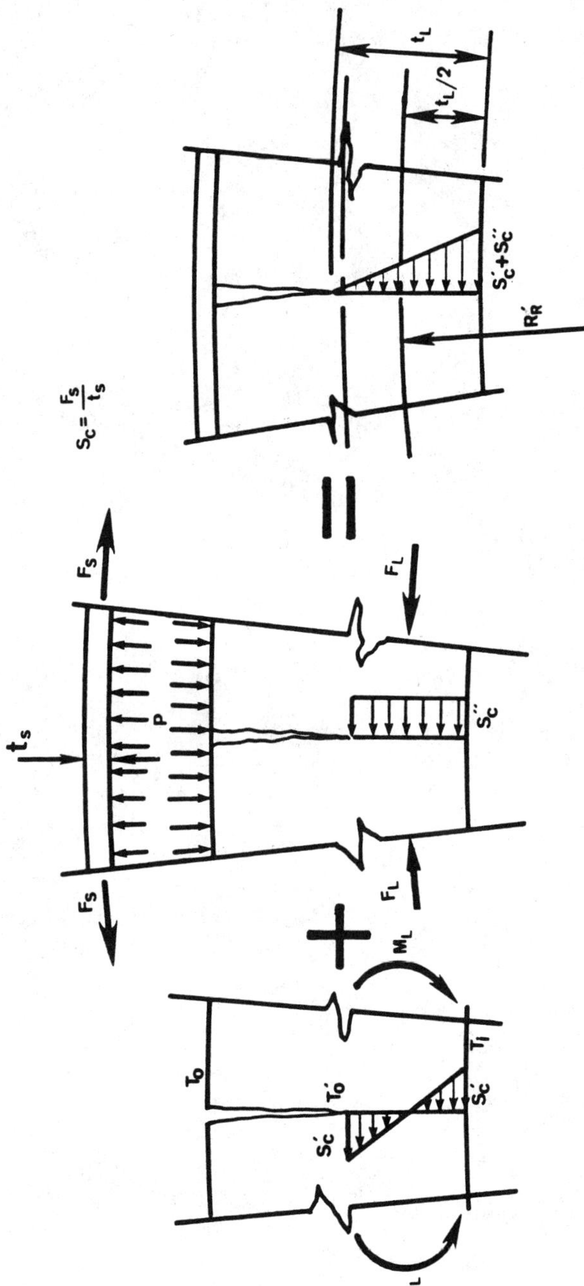

FIG. 2. Components of Expansion Stress in a Cylindrical Refractory Lined Vessel

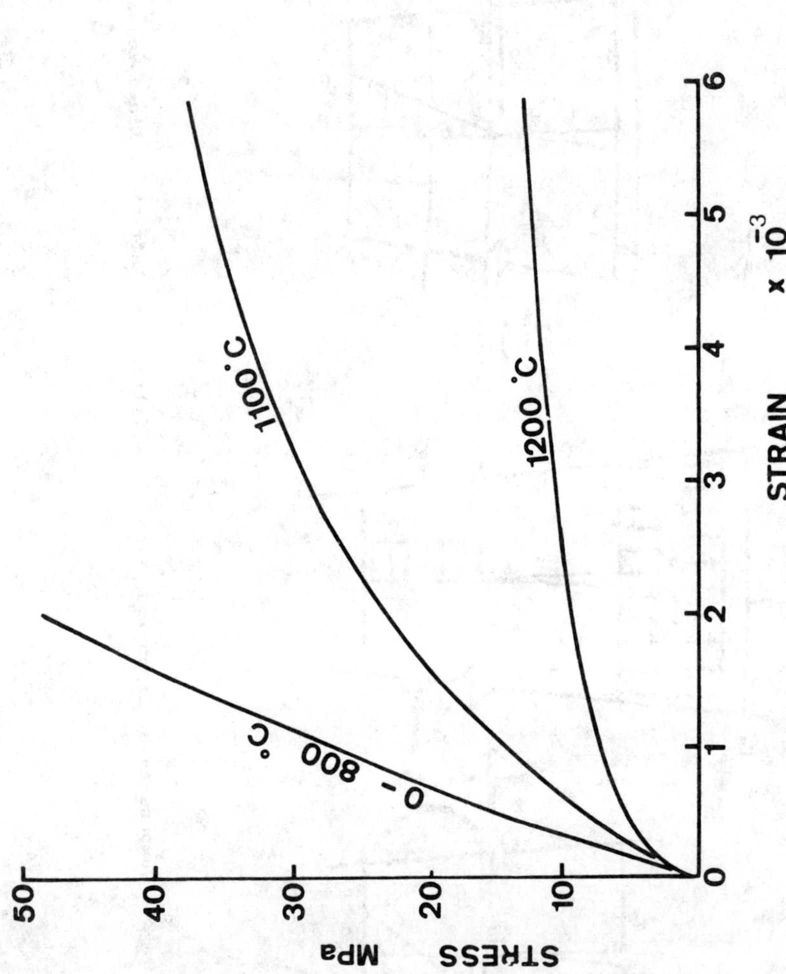

FIG. 3. Compressive Stress/Strain Data - 70% Alumina Brick

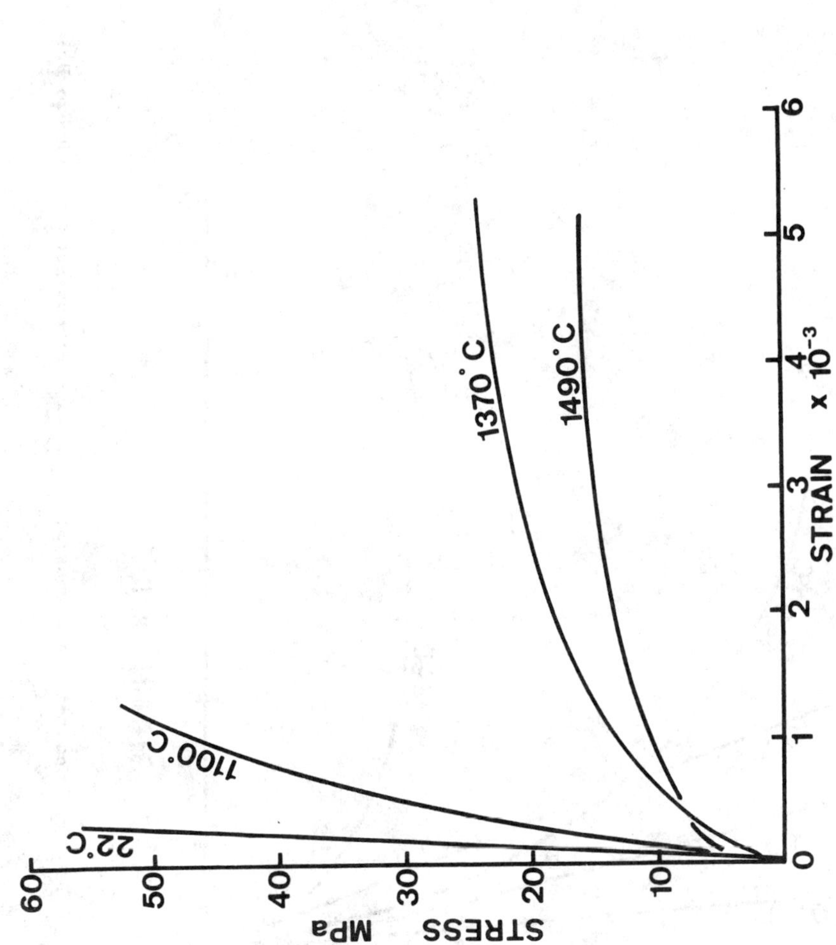

FIG. 4. Compressive Stress/Strain Data - 98% Alumina Castable

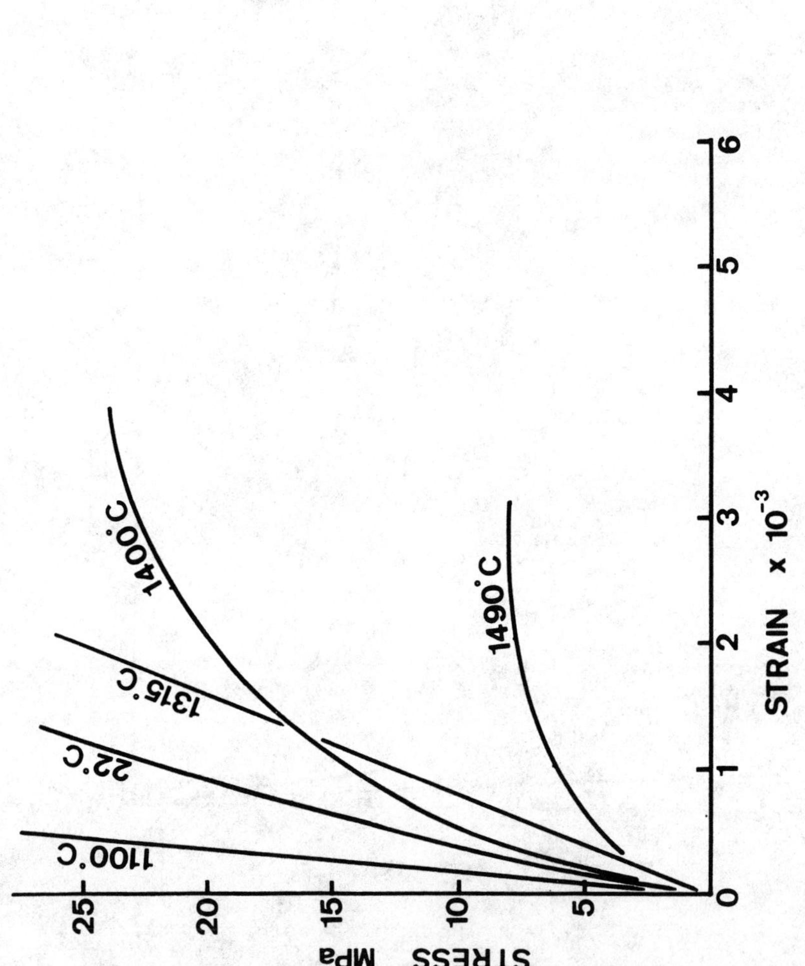

FIG. 5. Compressive Stress/Strain Data - Direct Bond Mag-Chrome Brick

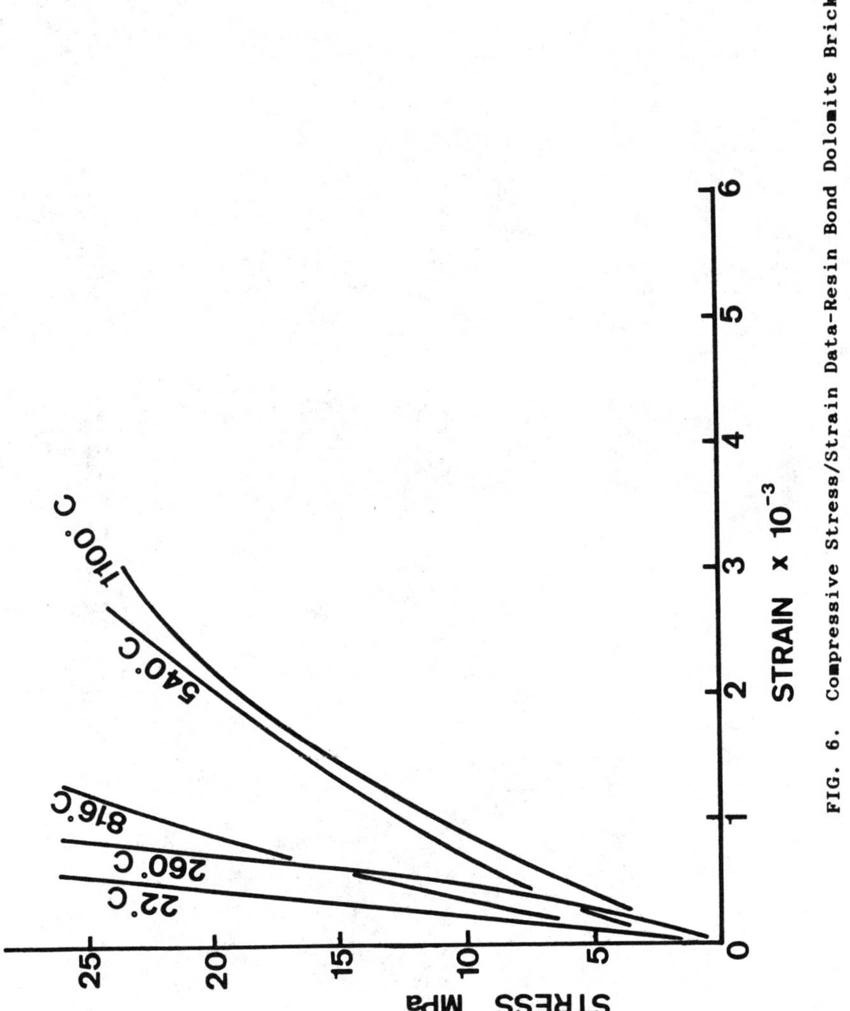

FIG. 6. Compressive Stress/Strain Data-Resin Bond Dolomite Brick

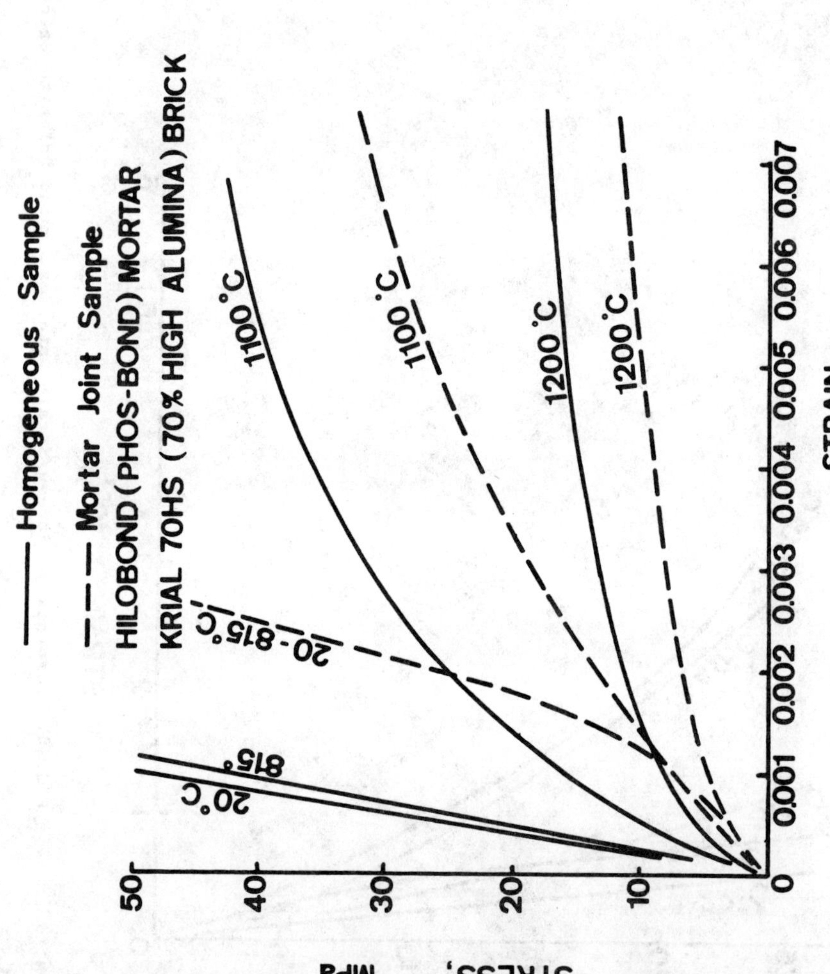

FIG. 7. Stress/Strain Curves for Brick/Mortar

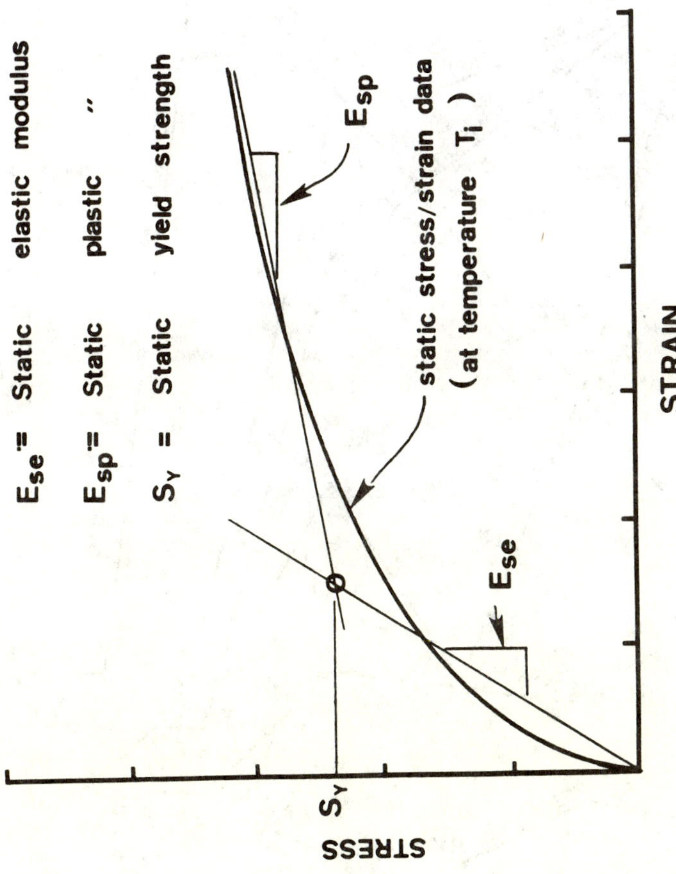

FIG. 8. Bilinear Curve Parameters

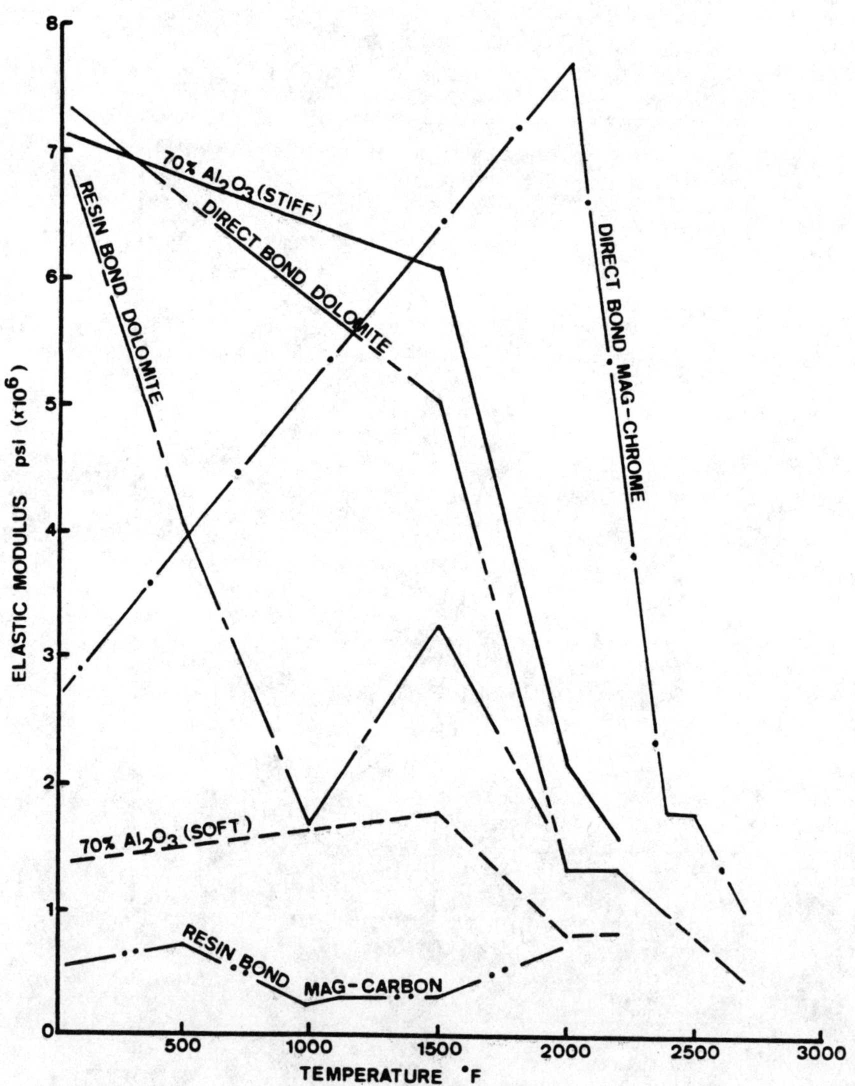

FIG. 9. Elastic Modulus of Refractory Brick

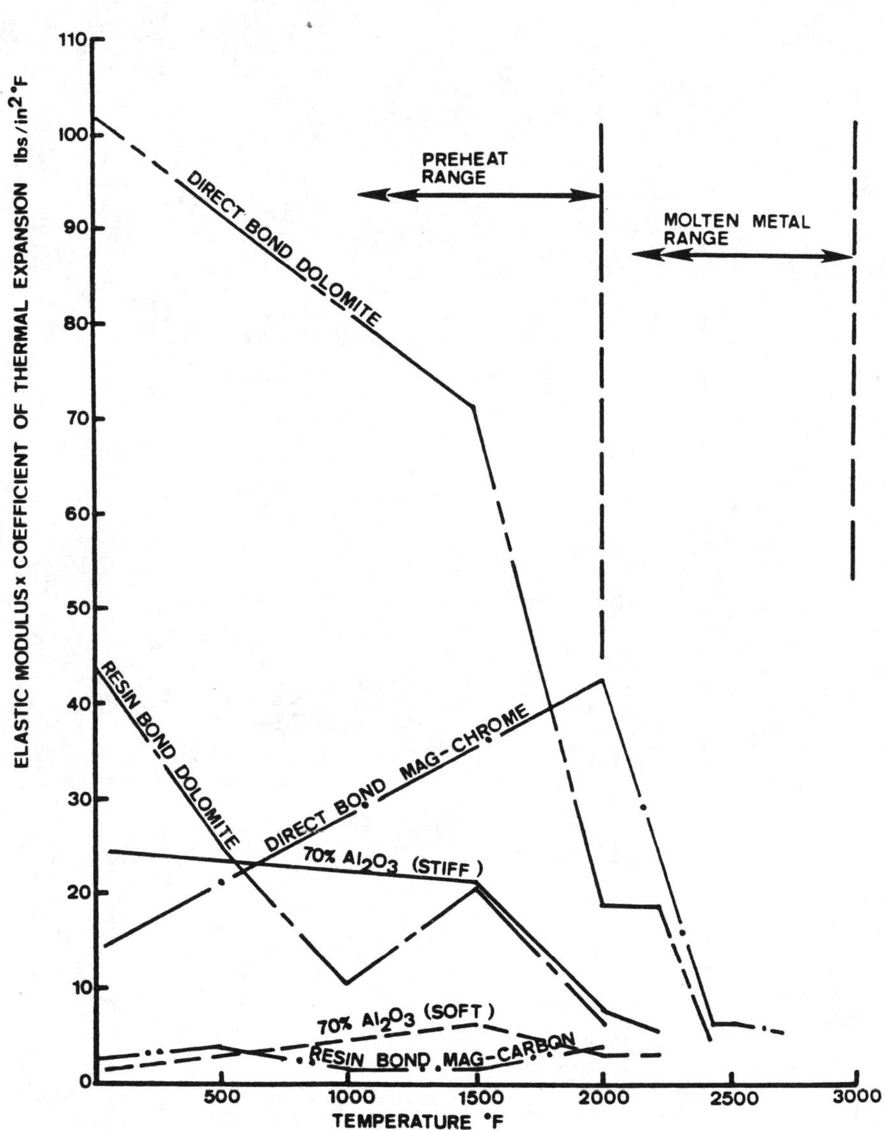

FIG. 10. Potential Expansion Force of Refractory Brick

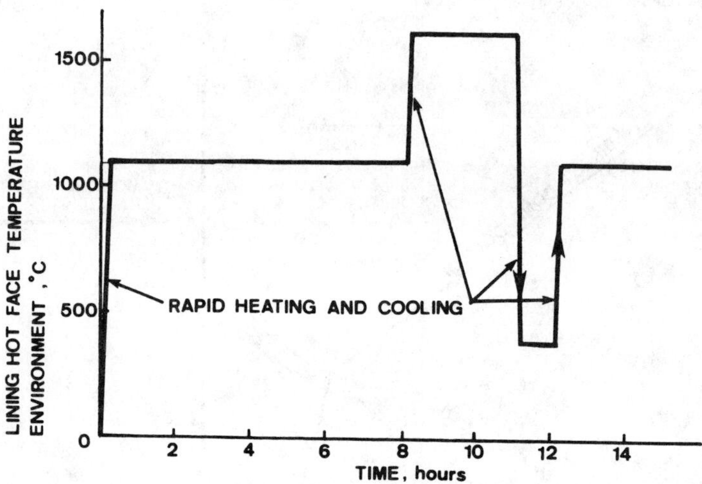

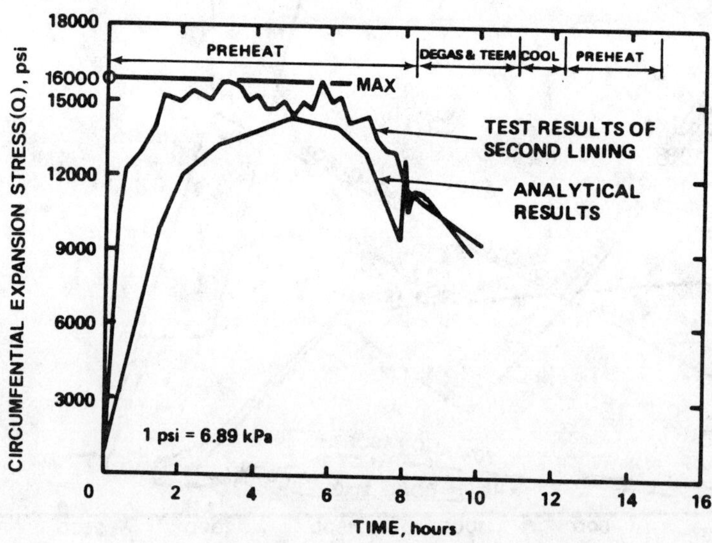

FIG. 11. Vessel Shell Expansion Stress (Lined with 70% Alumina Brick)

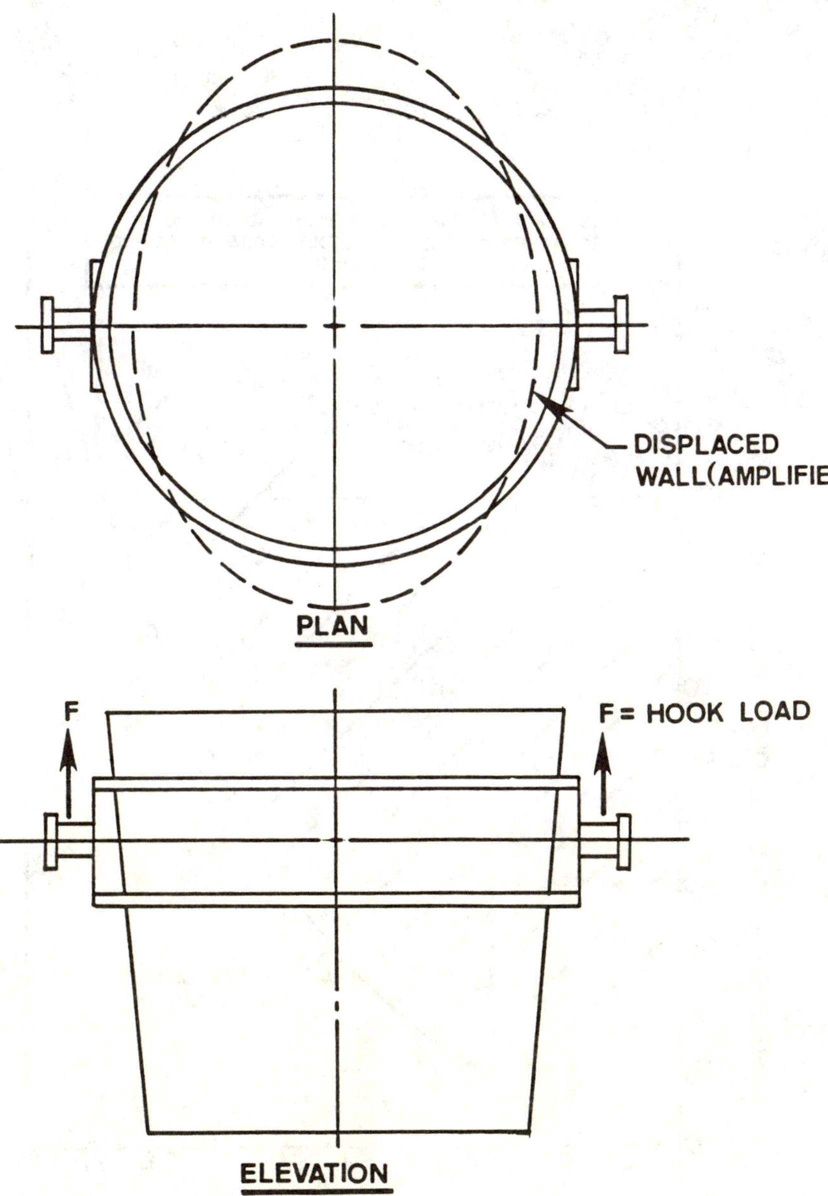

FIG. 12. Schematic of Ladle Wall Displacement During Lifting

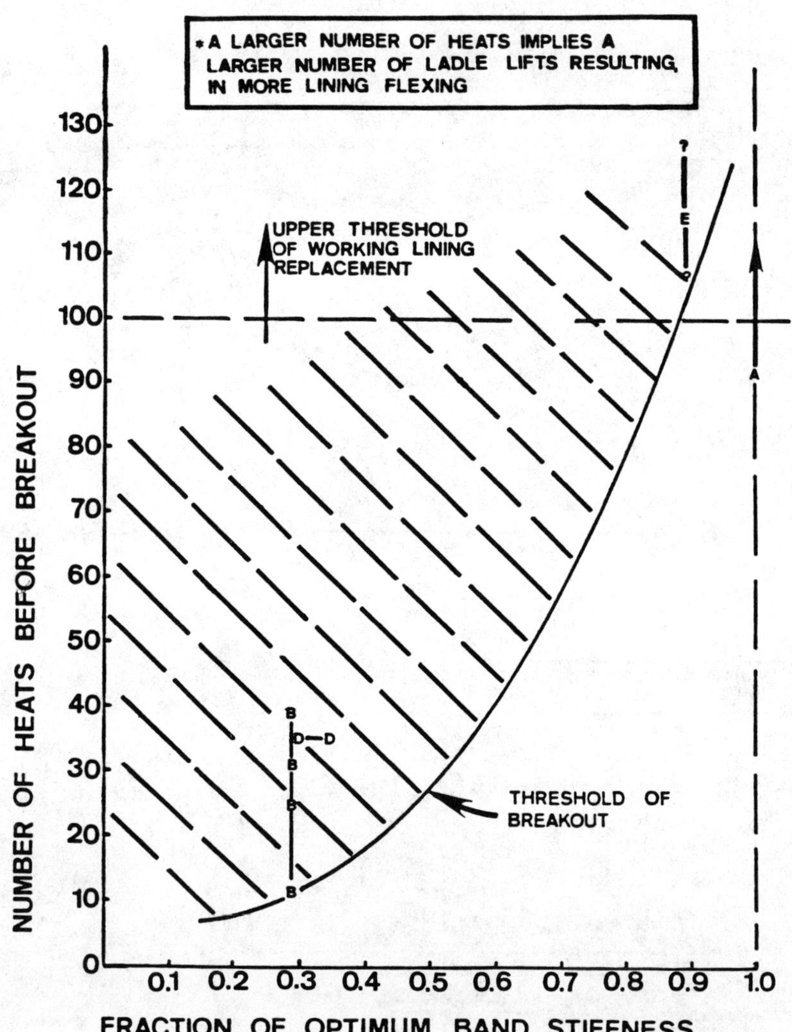

FIG. 13. Service Life of Brick in Ladles

Zinc ingot depalletizing and automated furnace charging

P. Garner and D. Bowskill
Aisco Systems Inc., Burlington, Ontario, Canada, L7R 3Y5

ABSTRACT

Automation of zinc ingot handling for furnace charging ensures high labour productivity and consistent furnace charges.

AISCO Systems Inc. engineered and manufactured a zinc ingot depalletizer, ingot elevator and furnace charging system for an Electrolytic Zinc Plant in Quebec. The system components included a slat conveyor to store up to 16 pallet loads of ingots, a depalletizer and ingot turnover machine, ingot elevator and distribution conveyors to charge two independent furnaces. System operation was tied into the plant master PLC and received instructions from furnace level indicators.

The system was material handling automation between processes. The same system can be modified to suit ingot depalletizing and furnace charging for hot dip galvanizing lines and die cast furnace charging.

INTRODUCTION

A Zinc Powder Plant, was constructed as an addition to an existing electrolytic zinc facility in Valleyfield, Quebec. The powder plant itself was automated, requiring only two operators to monitor its operation and package the product.

The plant required a combined depalletizing/furnace charging system that would respond automatically to a low level signal from either furnace. A forklift operator from the storage area would be used part time to keep the system supplied with bundles. Ingots were stored on ground level in bundles of forty. Due to the physical arrangement of the process, two melting furnaces were located in the upper level of the building. The ingots needed to be depalletized and elevated to the level where the melting furnaces were located.

In order to maintain consistent furnace temperature, charging was restricted to a maximum of one ingot every 60 seconds, per furnace. Another restriction would allow only five consecutive ingots at 60 second intervals, again to maintain consistent furnace temperature.

Fig. 1

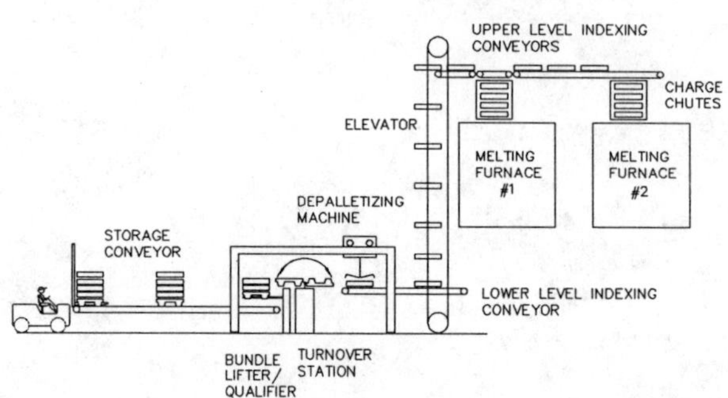

THE DEPALLETIZING & CHARGING SYSTEM - AN OVERVIEW

The system can be broken down into two areas - depalletizing and furnace charging. Refer to Figure 1 for a system overview. Charging is done on the upper furnace level. Single ingots are carried up on an indexing elevator. When an ingot is charged, the entire conveying system will index to fill the space.

The depalletizer is located on ground level, and cycles faster than the charging system. It focuses on keeping the first position on the lower indexing conveyor full after the charging system has indexed. Due to the pick and place nature of the depalletizing head, the bundle position is qualified to ensure consistent ingot placement. The bundle is lifted clear of the storage conveyor and into the qualifier above. Once qualified, it is lifted to the pick-up level, where suction cups on the depalletizing head can reach it. When a layer of ingots has been used, the next layer is qualified and lifted to the pick-up elevation.

The charging system is comprised of three indexing conveyors and two charge chutes. At the end of each charge chute, is a retractable stop and pusher. An ingot is held by the stop until the furnace demand signal is received. The stop then lowers, and the pusher charges the ingot into the furnace. The rest of the system then indexes to fill the space at the charge chute.

In order to maintain bundle stability, the layers are arranged in two different ways and alternated through the bundle height. (Refer to FIgure 2.) The bottom layer is comprised of ingots with feet, eliminating the requirement for skids. These ingots are inverted, to keep the "feet" up, for easer handling further down the process. This results in eight possible pick-up locations and two placement locations.

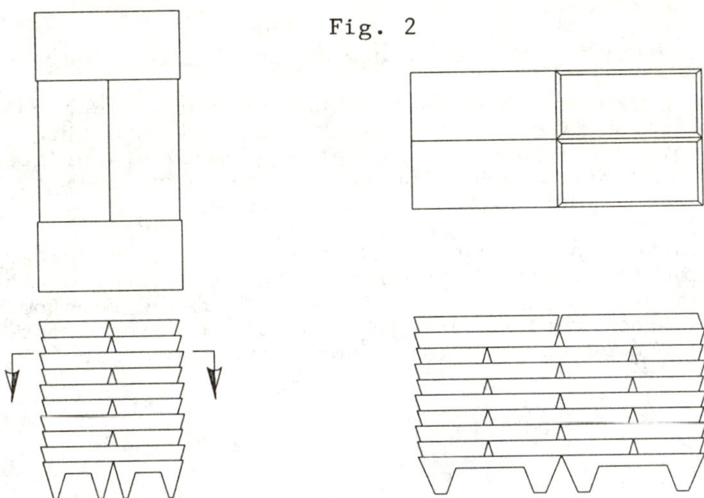
Fig. 2

Electrical control of the system utilized the plant Gould Modicon 984-680 CPU via a remote I/O interface and isolated I/O modules.

A requirement of the electrical design was that the system could be isolated into five zones. This feature enables limited maintenance to be carried out while the system is still running production.

Zone No. 1 Bundle storage conveyor and the depalletizer.

Zone No. 2 Lower indexing conveyor and elevator. If Zone No. 1 is isolated, ingots can be manually loaded on the lower indexing conveyor. The operator has control of the index via a local station push-button box.

Zone No. 3 Upper indexing Conveyor No. 1 and Conveyor No. 2. If Zone No. 2 is isolated ingots can be manually loaded and indexed via a local station pushbutton box located at upper indexing Conveyor No. 1 entry position.

Zone No. 4 Furnace No. 1 Feeder. If Zone No. 4 is isolated, ingots are fed to Zone No. 5 only.

Zone No. 5 Upper indexing Conveyor No. 3 and Furnace No. 2 Feeder. If Zone No. 5 is isolated, ingots are fed to Zone No. 4 only.

The ingot handling is monitored continuously by the PLC. If a delay is detected or anticipated, i.e., depalletizer working on last ingot bundle, lower indexing conveyor load position engaged for excessive time, an intermittent alarm will sound to alert the operator.

SYSTEM COMPONENTS

Bundle Storage Conveyor (Reference Figure No. 3)

The bundle storage conveyor is designed to convey 16 bundles totalling 18 tons. Carbon steel slats, bolted onto triple horizontal roller chain and driven at 75 fpm by a 4 HP gearmotor. The frame is a fabricated hollow structural steel design, with 3 ft. high side guides to contain the bundles.

A selector switch on a remote pendant enables the operator to put the bundle conveyor into a manual mode while loading of the bundles takes place. The automatic cycle for the remainder of the system is not interrupted during the bundle loading process. Even spacing of the bundles is obtained by using photo-electric sensors.

Fig. 3

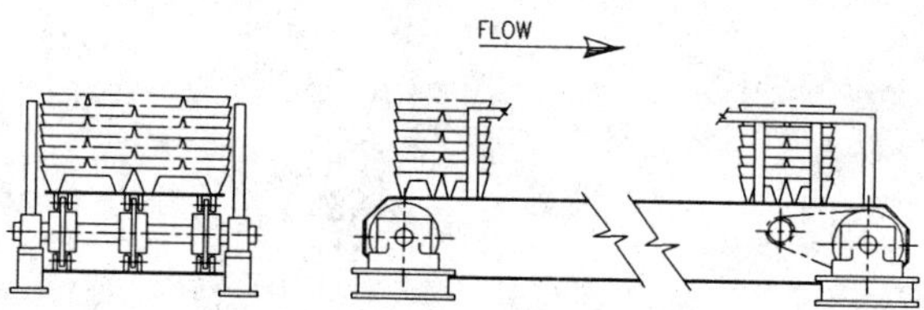

Bundle Lifter/Qualifier (Reference Figures No. 4 & No. 5)

The ingot bundle is carried over the lifter forks by the bundle conveyor. The forks are raised by a rack and pinion drive, and are guided by linear bearings. The bundle qualifier is pneumatically operated with four heavy duty cylinders. It squares up the top two layers of the bundle from all four sides to qualify the pick-up positions. Device positioning is obtained using proximity sensors.

Fig. 4 Fig. 5

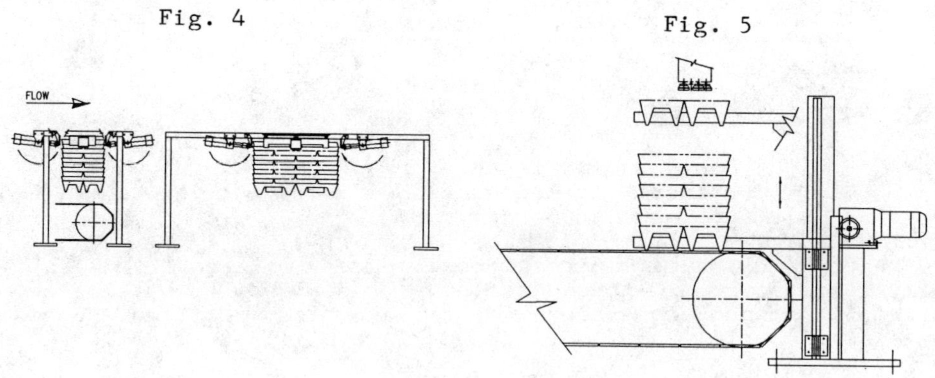

Depalletizer (Reference Figure No. 6)

The depalletizer unit has four degrees of freedom. The upper carriage and trolley provide the horizontal movement, and the suction head can raise, lower, and rotate. Four 2.5" diameter suction cups with two high flow rate vacuum generators are used to lift an individual ingot from the bundle. The high flow rate is required to compensate for cup leakage due to the uneven ingot surface. Toggle clamps secure the ingot before horizontal travel. The trolley and carriage are driven independently, by variable speed DC drives. All other motions are accomplished pneumatically.

Fig. 6

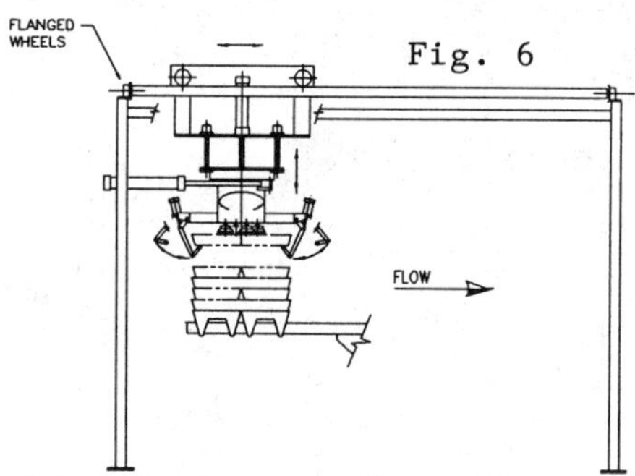

A small graphic display on the operator console indicates the ingot position the trolley/carriage is headed. If power is lost during cycle the next pick-up position will be restored on power-up. The operator, by using pushbuttons, can change the next pick-up position (if required). Positioning of the trolley/carriage over the ingot is achieved using inductive proximity switches. The proximity switch prior to the final destimation is used to slow the drive speed for increased position accuracy.

Turnover Station & Lower Indexing Conveyor
(Reference Figures No. 7 & No. 8)

The bottom layer of ingots must be inverted due to their feet (reference Figure No. 2) which cannot be accommodated by the conveyors. The ingot is deposited by the depalletizer and clamped pneumatically at each end. A reversing brake-motor drives the turnover and the ingot is deposited flat side down on the lower indexing conveyor.

Fig. 7

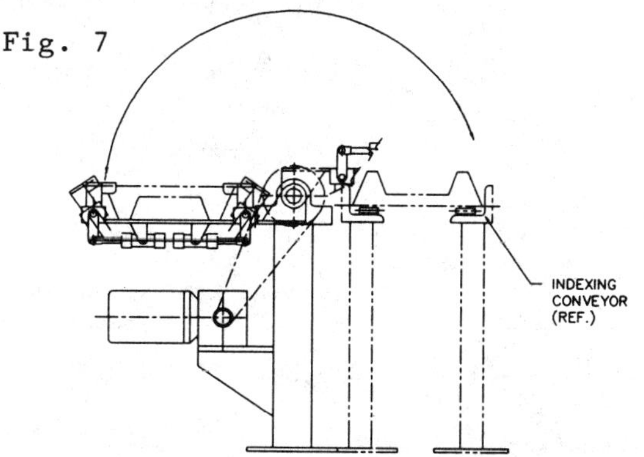

The lower indexing conveyor design is typical of all conveyors (reference Figure No. 8). It utilizes dual horizontal chains by a gearmotor with a pneumatic clutch and brake for accurate positioning. Like the other components in the system, the conveyors are a fabricated hollow structural steel design.

Fig. 8

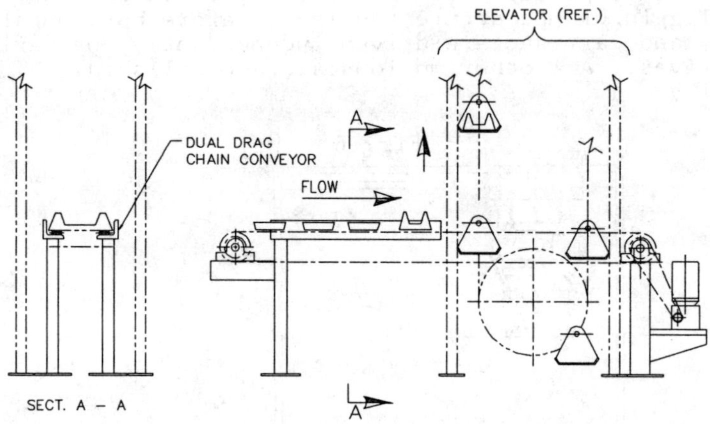

Ingot Elevator (Reference Figure No. 9)

The ingot elevator has an overall height of 44' and is assembled in three sections. It is chain type, with bolted on bucket attachments to carry the ingots. The attachments are guided by cam followers to keep the ingot horizontal as it travels over the head shaft. The drive, like the conveyors, is controlled with a pneumatic clutch and brake for accurate indexing. A back-stop is used as a safety in the event of brake failure. The frame is fabricated of hollow structural steel, and is guarded on all four sides, top to bottom.

Fig. 9

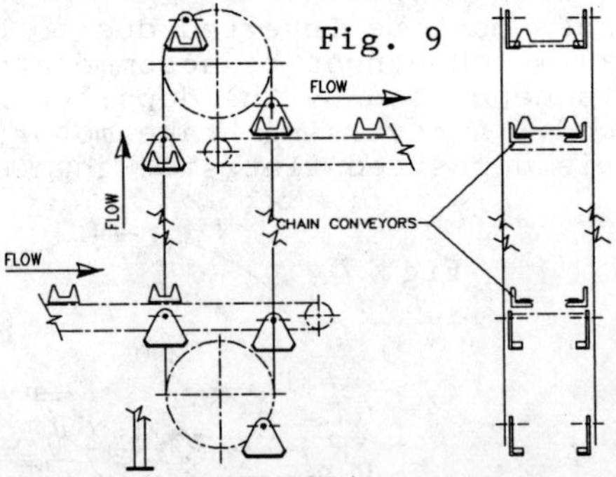

The elevator is interlocked with the lower indexing conveyor and the upper indexing Conveyor No. 1. When the drop off point on the upper indexing Conveyor No. 1 is vacant the elevator will cycle, providing the interlocking requirements are met.

Upper Conveyor Lifter & Feed Conveyor
(Reference Figures No. 10 & No. 11)

The ingots are lifted clear of the upper conveyor chains by a pneumatic lift. Once clear of the chain, the ingot rolls by gravity down the feed conveyors and is held by a pneumatic end stop. Ingots are pushed by a pneumatic cylinder which is guided by a Thompson shaft and bushing. Due to the heat radiation directly over the furnace, a stainless steel chute is used to guide the ingots into the furnace.

Fig. 10 Fig. 11

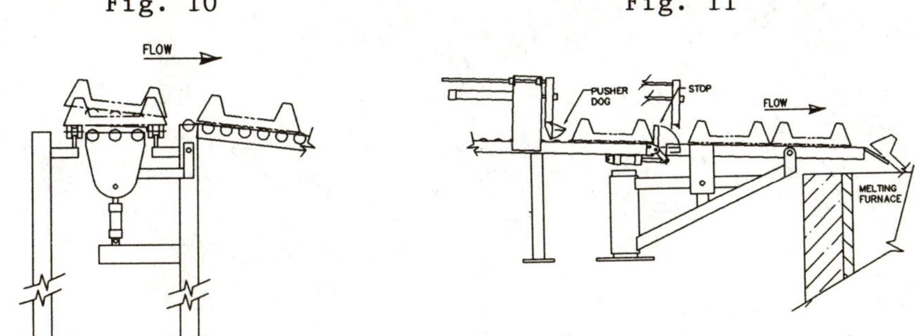

LONG TERM BENEFITS & OTHER APPLICATIONS

The major benefits of this automation are as follows:

* Low Operating Cost
* Consistent On Demand Charging

In production, the system requires only that the bundle conveyor be kept full. With the system operating to full capacity the bundle conveyor would need refilling every 5 hours. Due to the remote control pendant the operator can load bundles any time during automatic operation, giving him flexibility to do other tasks.

PLC controls results in consistent and on demand furnace charging. This minimizes furnace temperature and level fluctuations, and helps keep the powder making process running at peak efficiency.

The depalletizing and automated furnace charging system can be used in other applications. Galvanizing lines, continuous die casting, and aluminum billet charging could all benefit from this system. The depalletizer and fee conveyors can be modified to handle a different size ingot or bundle, and the material handling customized to suit any plant layout. In fact, the depalletizer could also function in reverse, as a palletizer.

This system is an excellent example of automated material handling in pyrometallurgy. It can realize long term benefits for many different processes and can be easily customized to accommodate each.

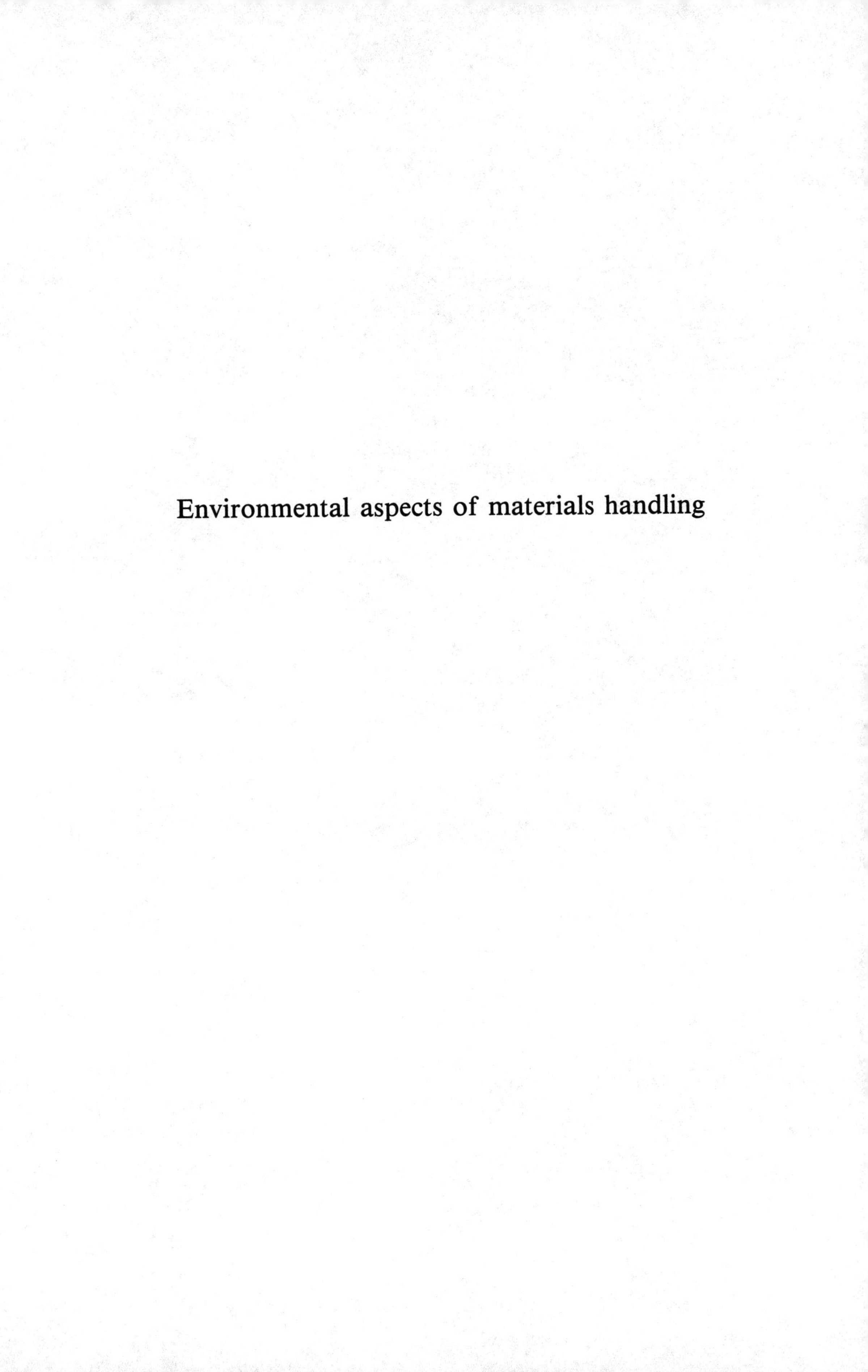

Environmental aspects of materials handling

A manganese mill dust explosion

J.A. Senecal
Fenwal Safety Systems, Inc., Marlboro, Massachusetts, 01752, U.S.A.

Introduction

In the spring of 1989 a manufacturer of metal powders experienced a process dust explosion in a manganese pulverizing tower during an unscheduled maintenance procedure. There were no personal injuries but equipment was damaged and down time was experienced. This paper reviews the essentials of the incident and summarizes the probable causality. In addition, explosibility and ignition characterization data is presented which aids in understanding the initiation and severity of the incident. As part of the analysis a review was made of the adequacy of the pressure relief vent design. Further, an assessment has been made of the feasibility of the use of deflagration suppression as a means of explosion protection in fine metal powder process systems. The result of the latter evaluation indicates that deflagration suppression may be a viable alternative, or complimentary, explosion protection method for metal dust systems.

The Process System

The firm, which experienced the incident that is the subject of this paper, prepares finely divided manganese metal powder which is combined with aluminum powder in one process for preparation of master alloys used in formulating manganese- hardened aluminum alloys. One application for these alloys is in the manufacture of sheet stock used to form beverage containers.

The method used to prepare manganese powder starts with chips of electrolytic manganese plate measuring 2 to 6 mm in thickness and 10 to 40 mm across. The Mn chips are fed to a ball mill located at the bottom of a comminuting tower, a free-standing structure located out doors well away from habitated or other process structures. Material discharged from the ball mill is picked up by a bucket elevator and conveyed vertically about 10 m where it is off loaded at the inlet of a rotary valve from which material is fed into a vibrating Rotex screen classifier. The functions of the rotary valve are to isolate the bucket elevator tower from the remainder of the system and to regulate the rate of powder feed to the classifier. The classifier was normally fitted with No. 30 (0.595 mm openings) screens. On the day of the explosion an evaluation was being conducted of the production of finer powder using No. 45 (0.354 mm openings) screens. Material which passed through the screens was passed to a duct through which it fell to a rotary valve which in turn fed the product to tote bins to which was later added the required amount of aluminum dust needed to prepare finished master alloy of the desired composition. Over-size manganese powder was passed to a down chute which returned material to the ball mill feed point for reprocessing. The down chute

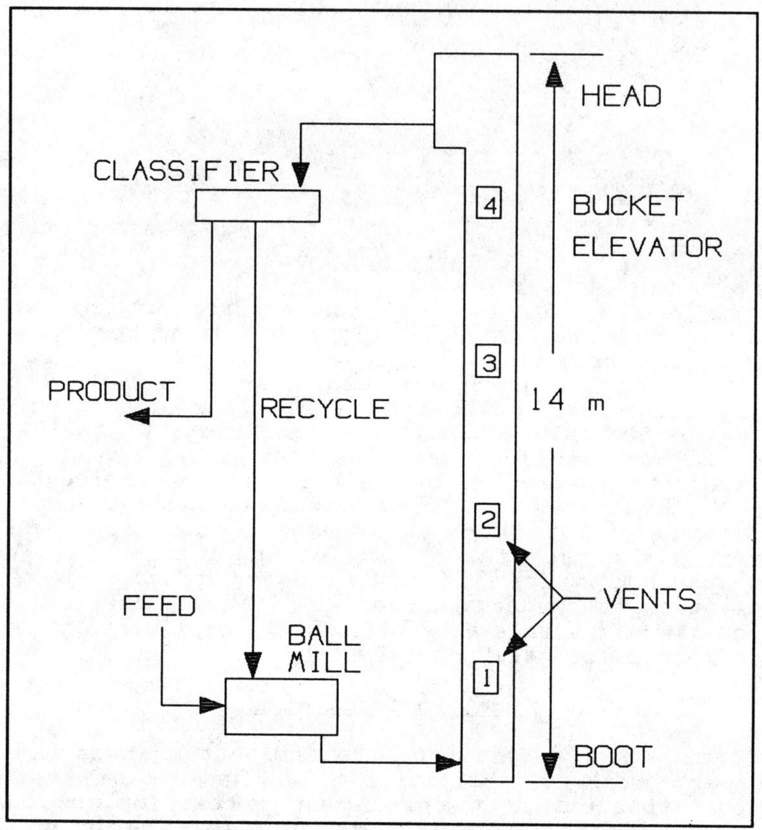

FIG. 1. Manganese milling system

consisted of a Linatex pipe, a rubber-lined flexible duct. The process is shown schematically in Fig. 1. A particle size analysis typical of normal production batches is given in Table 1.

The entire process, after the feeding of the Mn chips, is closed. The system is continuously purged with nitrogen to which sufficient air is added to maintain an oxygen concentration of 1 to 5 vol%. A low concentration of oxygen must be maintained in systems such as these in order to passivate the freshly formed metal surface created in the course of milling. Failure to do so would result in a powdered metal mass which was unstable with respect to atmospheric oxygen and would likely be pyrophoric. Oxygen analyzers were used to monitor O_2 concentrations at several locations in the process volume. These analyzers had alarm capability to warn of O_2 concentrations which were either too high or too low. All system components in contact with the Mn dust were bonded and grounded.

Employees working on or near the comminution tower were required to wear grounding straps on their shoes to avoid buildup of body charges. Non-sparking tools were used.

In addition to the measures taken to prevent accidental ignition of a manganese fire or dust explosion the bucket elevator was fitted with four pressure relief panels measuring 0.46 x 0.61 m. The panels were designed to relieve at an internal static pressure of 21 kPa g. The placement of the vent panels on the bucket elevator is illustrated in Fig. 1.

TABLE 1

Typical Manganese Powder Sieve Analysis

Screen Size (mesh)	Screen Openings (mm)	Amount On Screen (wt %)
20	0.841	0.0
40	0.420	8.0
100	0.149	49.2
200	0.074	16.0
325	0.044	8.1
Pan	<0.044	26.7

The Incident

Operation of the comminution tower was generally without incident. On the day of the event the classifier screen deck became plugged with a large quantity of what turned out to be very fine powder, i.e. smaller than 200 mesh (0.074 mm). This could only have occurred if the screen deck had not been installed properly or if it had become dislodged during operations.

On discovery of the problem the comminuting tower was shut down. A two-man crew and a supervisor proceeded to evaluate the problem. Before proceeding to open the classifier the work crew shut off power to the comminuting tower operating units by throwing an emergency switch located near the classifier. It was later acknowledged that this was not a proper shut down procedure particularly in as much as **all** operations on the tower were shut off buy this switch including the oxygen monitoring and nitrogen purge systems. The latter point was not understood or appreciated at the time of the incident. The cover of the classifier was removed which revealed the large backup of powder and showed that the 8" Linatex duct was clear. Aluminum shovels were used to remove the powder accumulated on the classifier screen deck and deposit it into the inlet of the down chute.

During the clearing of the clogged classifier an explosion occurred. A flame ball and a cloud of brownish-orange dust were ejected from the inlet of the ball mill return chute into which the work crew had been shoveling manganese dust. Two of the four pressure relief vents on the bucket elevator were ruptured, one completely and one partially. Dust and fire balls issued forth from opened vents. The workmen were able to step back away from the classifier to a rail. They were temporarily enveloped in a cloud of dust. The work crew was able to immediately escape the work deck by means of a nearby stair way. No one was injured in the incident either by the fire flash or by dust exposure.

Even though the bucket elevator was protected by pressure relief panels mechanical damage was sustained by the top section of the elevator. This section of the elevator was constructed of 12 gauge (2.77 mm) sheet steel. The cost of replacing the vent panels and repairing the damage to the elevator amounted to about $2500, a relatively minor cost considering the potential damages. For example, the mill was out of operation for 24 hr loosing production of 5,400 kg of manganese powder. However, this powder and its associated aluminum master alloy were not on back order. Had this been the case the cost in lost production time would have been approximately $145,000. The costs associated with the potential injury of one of the crew are difficult to estimate but would certainly have been appreciable. This is not to say, of course, that dollar costs fully reflect the seriousness of injuries to personnel.

Analysis

Manganese Combustibility

As a first step in analyzing the manganese explosion a determination was made of the combustibility characteristics of manganese dust. Data are given in Table 2 which characterize the sensitivity to ignition and explosibility of manganese dust. Data, determined by the same methods, are also given for Pittsburgh seam coal dust, to serve as a frame of reference. The explosion potential of coal dust is well appreciated and has been studied at length by the U.S. Bureau of Mines.

The data indicate that manganese dust is explosible at a low dust cloud concentration, at only about 30% that of coal dust on a particle volume basis. (The specific gravities of manganese and coal dust are 7.2 and 1.35, respectively.) The ease of ignition of manganese as a dust cloud is somewhat more difficult than for coal dust based on the dust cloud ignition data. However, the manganese dust cloud ignition energy and temperature values are not high

TABLE 2

Ignition and Explosibility Characteristics of Manganese Dust

	Manganese	(Reference) Pittsburgh Coal Dust
Minimum Explosible Concentration (g/m^3)	250	120
" " " ($m^3/m^3 \times 10^6$)	35	89
Minimum Spark Ignition Energy (mJ)	460*	160
Ignition Temperature Dust Cloud (°C)	454*	510
Ignition Temperature Dust Layer (°C)	223*	165
Minimum Oxygen Concentration to Support Combustion (Vol %)	14	18
Maximum Deflagration Pressure (kPa g.)	670	810
K_{st} (MPa-m/s)	15	12

Notes: 1. Data are for samples passing a 200 mesh screen.
2. * data are for samples passing a 325 mesh screen.

enough to preclude electrostatic or mechanical spark ignition. Metal powders with oxidized surfaces are efficient capacitors (1) and substantial discharge energy potential exists. Dust layer ignition occurs at relatively modest temperatures, values of which could be achieved due to self heating by oxygen adsorption. The minimum oxygen necessary to support deflagration of a dust cloud is relatively low at 14 vol per cent. This data does not to imply that dust **layer** combustion will not be supported at lower oxygen concentrations.

Dust cloud explosibility is characterized by the maximum deflagration pressure attainable and by the flame speed which the dust cloud is capable of supporting. Dust cloud flame speed manifests itself in the maximum rate of pressure rise, $(dP/dt)_{MAX}$, attained in closed vessel deflagration tests. A standard method (2) is employed to determine $(dP/dt)_{MAX}$. The value of this term is dependent on the size of the test vessel. This term may is effectively normalized as:

$$K_{st} = (dP/dt)_{MAX} V^{1/3} \qquad (1)$$

where V is the volume of the test vessel. Once determined K_{st} may be used to predict rates of pressure rise in larger process volumes, design pressure relief vents (3) and is an important parameter for the application of a proprietary design method for deflagration suppression systems.

The maximum deflagration pressure and K_{st} values are similar to those of high-grade coal dust. These results can be attained by manganese only by very rapid oxidation rates and high flame temperatures. The latter point can be appreciated by the observation that deflagrations of manganese and coal dust achieve similar maximum pressures even though manganese oxidation reduces the net number of moles in the gas phase by 21% due to oxygen abstraction to form metal oxide while coal combustion achieves practically no net change in the number of moles of gas after reaction.

<u>Incident Review</u>

The mechanism leading to the initiation of the manganese dust explosion can not be known with certainty. However, the most probable causes of this event were deduced from an analysis of the available data. The following factors seem to have been very important in establishing the necessary conditions for initiating the explosion:

a. Material backup at the operating ball mill resulted in excessive grinding of manganese to a mean particle size probably much smaller than usual. The resultant increase in specific surface area would have caused this metal dust to be more easily ignitable and more violently explosible than product having a more normal particle size distribution, given in Table 1.

b. The inadvertent shut off of the nitrogen purge system, as noted above, allowed the oxygen concentration within the system to rise to levels able to support combustion.

c. The process of shoveling manganese dust from the clogged classifier screen deck and dumping it down the Linatex return duct had the effect of "pumping" air down into the bowels of the system, and in particular elevated the oxygen concentration at the ball mill feed recycle inlet.

d. A blind flange, which would normally have been inserted at the bottom of the Linatex return duct during scheduled maintenance operations was not put in place during this unscheduled procedure. This flange would have prevented air from being delivered to the ball mill area as described in point (c).

e. It was observed that the rubber liner of the Linatex duct was not fitted with an interior conductive element between its termini. Metal dust sliding down the rubber surface would very likely undergo significant triboelectric charging. Charged powder would eventually discharge when it reached a location of suitably lower potential.

Thus, the four conditions necessary for an explosion were met during this unscheduled maintenance procedure:

1. Fuel - An ample supply of easily ignitable manganese dust existed throughout the process volume, and in particular, at the ball mill location where finer-than-usual dust had been created.

2. Oxidant - Oxygen concentrations were rising in the whole process volume due to disabling of the nitrogen purge system and the circulation of air into the system by the actions of the crew.

3. Ignition Energy - Several possibilities exist which I have rated as follows:

a. Self heating (70%) - Adsorption of now-available oxygen by the very finely ground powder at the ball mill location would result in surface oxidation with a consequent release of heat. The thermochemistry of manganese oxidation is as follows (4):

$$Mn + 1/2\ O_2 = MnO + 385\ kJ \quad (2)$$

$$MnO + 1/2\ O_2 = MnO_2 + 136\ kJ \quad (3)$$

Thus, given the large heats of reaction involved only a modest degree of oxygen adsorption within a layer of fine dust, which is a very poor thermal conductor, is required to raise the material temperature to the point of smoldering or flaming.

b. Electrostatic discharge (25%) - Powder, charged in transit down the rubber lined duct, discharging at a ground point.

c. Frictional heating (3%) - The rubbing of manganese dust between shovel and fixed surfaces may have initiated localized smoldering of particles which were subsequently dumped down the Linatex duct.

d. Mechanical sparks (2%) - Sparks due to contact of shovel and fixed surfaces producing hot sparks.

Mechanism (a) seems to be the most likely cause of ignition in this instance as the requirements for self heating were well established. The conditions necessary to develop significant electrostatic charging (b) were also present. However, relatively high ignition energies are required to ignite dusts, even metal dusts, as data given below indicate. However, given the high proportion of dust smaller than 44 μm (see Table 1) the ignition energy data obtained on "similar" material may not fully reflect the sensitivity of this dust to electrostatic ignition. Frictional (c) and mechanical (d) sparking seem to be less likely ignition mechanisms due to the use of non-sparking tools and the experience of the crew in handling powdered manganese.

4. Confinement - Conditions 1, 2 and 3 above are sufficient to begin combustion. It is the relative enclosure of the combustion process which gives rise to development of excessive pressure. The manganese comminuting system was, by its very design, a nearly closed system. The only access to the bucket elevator volume was through the ball mill itself and the discharge rotary valve at the top of the tower which at the time of the incident was clogged with powder.

Review of Vent Design

Given that the causes of the event are now understood it remains to be determined why mechanical damage occurred in spite of use of pressure relief panels. The elevator has a height of 14 m and a cross section measuring approximately 1 x 1 m. The internal volume is about 14 m^3. The appropriate design guideline for pressure relief vents is NFPA-68 (3). Key points of the guideline relating to the protection of elongated enclosures are:

1. Total vent area at each vent location should equal the cross sectional area of the enclosure. NFPA 68 Sec. 8-3.2

2. Design for enclosures containing obstacles (buckets, belt, struts) and designed for a P_{red} of greater than 20 kPa g. calls for additional vent area. NFPA 68 Sec. 8-3.5

3. NFPA 68 Sec. 8-3.5 also acknowledges uncertainties with respect to the design of vents for elongated enclosures for dust systems.

4. The release pressure, P_{stat}, should be a low as possible consistent with equipment operation but in any case should not exceed one half of the design value of P_{red}. NFPA 68 Sec. 8-3.7.

The actual vent design used in the bucket elevator included four 0.28 m^2 pressure relief vents placed on 3 m centers and located as shown in Fig. 1 . The static relief pressure of each vent was 21 kPa g. Given the points of 1, 2 and 3 above it appears that additional vent area should be provided at each vent position. Both larger vents and vent placement on opposite sides of the elevator should be employed. Point 4 suggests that vent panels with lower relief pressure should be used.

Inspection of the vents after the explosion showed that vent 1 was not damaged; vent 2 was slightly bulged; vent 3 relieved; vent 4 was blown completely out giving the appearance of a violent event. The top section was bulged and internal angle iron was bent from 90 to 180 degrees. These observations suggest an accelerating dust cloud deflagration progressed up the elevator. Pressure in the head section probably exceeded 40 kPa g., twice the vent relief pressure, as evidenced by the structural damage. Given these results it seems that the recommendations of the previous paragraph are warranted.

Modification of the pressure relief design would reduce the probability of the development of a violent event in the future. It would not, however, mitigate the unavoidable consequence of ejection of flame, flammable dust and hot metal oxide dust into the environment. An alternative explosion protection method which is just now being evaluated for use with metal dust systems is explosion suppression.

Deflagration Suppression In Metal Dust Systems

Deflagration suppression is an explosion protection method which works on the principle of early detection of a developing deflagration followed by rapid delivery of an extinguishing agent, or suppressant, which quenches the growing fire ball. If the combustion process is not too rapid then flame extinguishment can be effected before the pressure rise exceeds the strength of the containing vessel. Deflagration suppression has been used to protect industrial process systems for over 30 years. However, the application of explosion suppression technology has been limited to systems where the explosion threat was due to the processing of combustible gases or carbonaceous dusts. Deflagration suppression has not been applied to metal dust systems to this point due to several fundamental differences relative to the nature of the explosion threat. Due to the development of new extinguishing agents and agent delivery systems it appears that substantial mitigation of metal dust explosion threats is now possible through use of deflagration suppression methods.

The principles of deflagration suppression as applied to systems processing carbonaceous materials have been well documented in the literature (5, 6, 7). They are applicable primarily to closed or semi-closed process volumes. Early detection of a developing deflagration in dust processing systems is normally achieved using pressure detectors, although rate-of-pressure-rise detectors are employed in certain cases. The detection threshold is set to as low a pressure as practicable, typically 3.5 to 7 kPa above the maximum excursions of the process pressure. A rise in pressure to the detection threshold is presumed to be due to a developing incipient deflagration. At detection of a deflagration the pressure detector makes a contact closure which completes a circuit through a power supply and the explosive actuators of the high-rate-discharge (HRD) extinguishers. Each HRD is filled with suppressant and pressurized with nitrogen. The suppressant is released into the protected space when a burst disk is caused to open by the action of the actuator. Extinguishment of the developing flame ball is accomplished primarily by thermal quenching mechanisms. Chemical kinetic mechanisms are also important in the case of certain agent-combustible combinations.

The reasons for which deflagration suppression has not been applied to metal dust systems to date are primarily:

a. Unavailability of suitable suppressants which were adequate to the task of extinguishing very high temperature metal-air deflagrations. The family of traditional suppressants consists of water, Halons 1301, 1011, 2402 and 1211. These agents are not generally deemed satisfactory in metal dust environments for reasons relating to effectiveness and materials compatibility. Water, for example can react with metal dusts to form hydrogen under some circumstances. However, two dry-chemical suppressants, consisting of specially formulated sodium bicarbonate and monoammonium phosphate, have been found to be particularly effective and especially so against metal dust deflagrations.

b. Unavailability of hardware which could deliver the agent at rates sufficiently high to overcome the "thermal momentum" of these systems which is a combination of high flame temperatures and high burning velocities. This limitation has been substantially overcome by use of improved-design HRDs and agent distribution nozzles.

The applicability of deflagration suppression to the protection of manganese powder processing systems has been evaluated. This was accomplished through a series of tests which first evaluated the ignitability and explosivity characteristics of manganese powder. The results of these tests are given in Table 2.

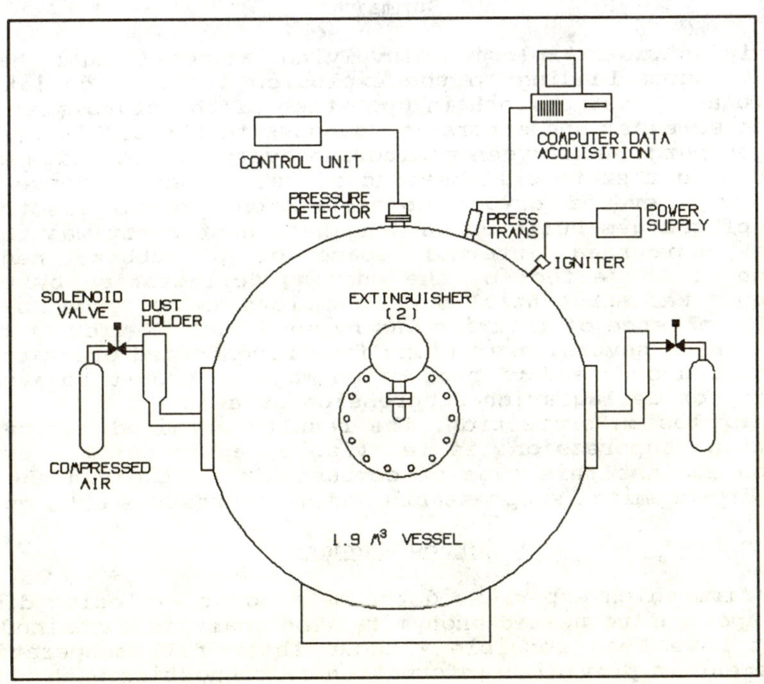

FIG. 2. 1.9 m^3 test vessel

The suppressibility of a manganese deflagration was next evaluated. The suppression test system is shown in Fig. 2. The spherical test vessel had a volume of 1.9 m^3 and was fitted with two dust dispersion systems, two 4 kg HRDs, a pressure detector and suppression system control unit. Vessel pressure was measures using strain gage pressure transducers and a digital data acquisition system. Each HRD contained 4 kg of Dessikarb (sodium bicarbonate based extinguishant) and was pressurized to 3.5 MPa g. with nitrogen. The pressure detector was calibrated to make contact closure at 3.5 kPa g. The test vessel was pre-evacuated to -4 kPa g. to compensate for the pressure rise due to the dust dispersion air. Manganese deflagration and suppression tests were conducted using sufficient manganese to achieve a dust cloud concentration of 2 kg/m^3. The results of these tests are shown graphically in Figs. 3 and 4. The maximum deflagration pressure was 670 kPa g. The maximum pressure obtained in the suppressed deflagration was 48 kPa g. A portion of that pressure rise, 14 kPa, was due to the nitrogen liberated from the HRDs. Thus, the net pressure rise due to the deflagrating manganese dust cloud was 34 kPa g. which is only 5% of the unsuppressed deflagration pressure.

Suppressions of deflagrations of carbonaceous dusts of similar K_{st} value typically achieve reduced pressures of about 20 to 30 kPa g. using 50% as much suppressant. The results achieved with deflagrating manganese dust reflects the difficulties associated with high metal dust flame temperatures.

Summary

An industrial explosion involving a metal dust has been reviewed. Events leading to the explosion included deviations in standard plant practice both in operation of the classifier (use of extra fine screens) and errors in response to the problem (shut off of nitrogen purge and oxygen monitoring systems). Ignition was most probably due either to self heating of very fine manganese dust on exposure to elevated oxygen concentrations or to electrostatic sparking of charges built up on manganese dust as it was shovelled down the ungrounded internal space of a rubber-lined duct. Mitigation of the effect of the ensuing deflagration by pressure relief vents was substantial but incomplete as evidenced by system damage. The absence of injuries was by good luck. Review of the vent design revealed several deviations from recommended practice. Tests were conducted on standard production manganese dust to assess the suitability of deflagration suppression as a means of explosion protection. The results obtained indicate that deflagration suppression is a viable approach to explosion protection against this type of combustible dust given the use of suitable dry-chemical suppressants and high-pressure extinguishers.

Acknowledgment

The firm which experienced the metal dust explosion discussed in this paper has requested anonymity. The analysis contained herein would not have been possible without their full cooperation and encouragement in providing information in connection with the event and manganese dust for conducting the experimental work on deflagration suppression analysis.

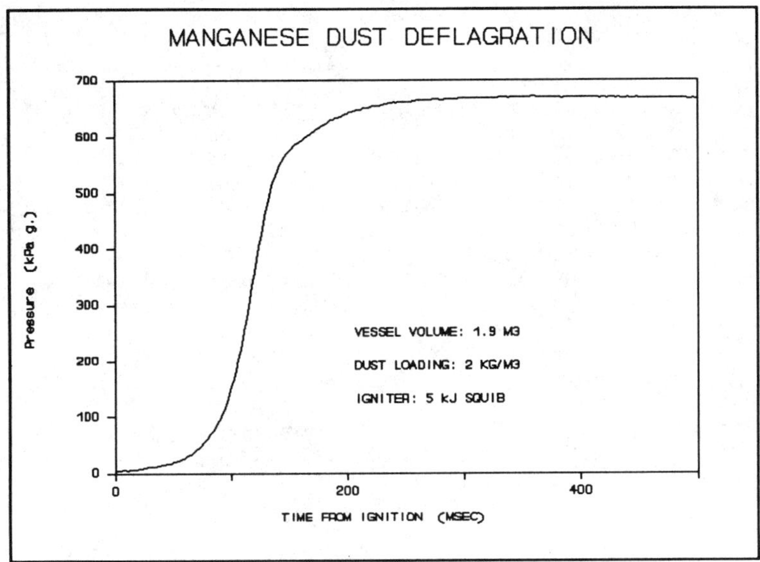

FIG. 3. Manganese dust cloud deflagration

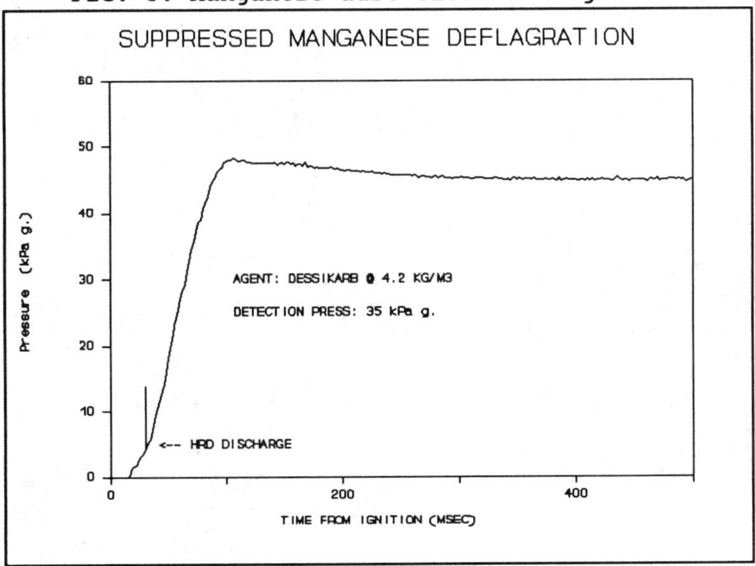

FIG. 4. Suppressed manganese dust cloud deflagration

References

1. Glor, M., Electrostatic Hazards in Powder Handling, pp.24-28, Research Studies Press, Ltd., Hertfordshire, England (1988).
2. Perry, J.H., ed., Chemical Engineer's Handbook, 4th ed., Table 3-201, McGraw-Hill, New York (1963).
3. NFPA 68, "Guide for Venting Deflagrations," National Fire Protection Association, Quincy, MA (1988).
4. "Standard Test Method for Pressure and Rate of Pressure Rise for Combustible Dusts," American Society for Testing Materials, No. E1226-88, Philadelphia, (1988).
5. Moore, P.E.,"Industrial Dust Explosions," K.L. Cashdollar and M. Hertzberg, eds., ASTM Special Publication 958, American Society for Testing MAterials, Philadelphia (1987).
6. Senecal, J.A., "Deflagration Suppression of High K_{st} Dusts," Plant/Operations Progress, Vol. 8, No. 3, pp.147-151, July (1989).
7. Bartknecht, W., Dust Explosions: Course, Prevention, Protection, pp. 203-214, Springer-Verlag, Berlin (1989).

The price of DURT — what fugitive materials cost conveyors and how improved systems can save money

R.P. (Dick) Stahura
Martin Engineering Company, Neponset, Illinois, U.S.A.

INTRODUCTION

My all-time favorite garage mechanic, gadgeteer, mechanical wizard, and philosopher lived in my home town in the coal fields of Western Pennsylvania. One of his businesses was a tire dealership, and so people would consult him on tire problems ranging from off-road earth hauling equipment to racing cars. One evening I got into a conversation with him about tires and I asked him for advice on how the average car-owner could prolong the life of his tires.

His answer was simple: "Wash your car daily."

His explanation was that when the average person has a clean car, he or she drives with more care. They ease out of the driveway. When following, they don't get too close, for fear of getting splashed. They don't want bug spatters all over the windshield, so they drive a little slower. They ease over railroad crossings. They come to a gravel road or to a construction zone and go much more cautiously.

"You watch the same driver when his car is dirty, and he bounces over railroad crossings, dirt roads, pot holes. He'll start faster, drive faster, and stop faster. In general, he abuses his machine. And so tires--and other components--just don't wear as well or last as long."

This conversation took place over 40 years ago. Now, from my own observations, I'd say he was right on target. And now, when a client asks me how they can improve profitability, my answer is: "Keep your operation clean."

At a clean plant, the equipment lasts longer and there is less unscheduled downtime for repair or replacement. So production efficiency stays up. In addition, the clean plant is safer, has better morale, and is hassled less by regulatory agencies.

Conversely, at a dirty plant, equipment breaks down, so it needs premature replacement with the added overhead of downtime for replacement. Employees reflect a sloppy attitude in both safety and in their attention to work requirements. And the regulatory agencies pay more attention. These factors can be considered "The Price of DURT." For clarification, let me give a definition of the word "DURT". It's my name for the material that escapes from the conveying system. It's the fugitive material underneath conveyor belts and around transfer points. It's the dust particles that collect around a conveying system. It's the wet muck that cakes on the support hardware of conveyors.

Let me review the reasons for my answer. A clean plant's equipment lasts longer. There are less surprise failures. Therefore, production that is scheduled is realized. Downtime--and the resulting loss of production--occurs more often in dirty plants.

THE BENEFITS OF A CLEAN PLANT

A clean plant handles a product as few times as is necessary, so efficiency is high. When material becomes fugitive and has to be gathered up and reintroduced into the system, this inefficiency increases the cost of production. Or if the fugitive material is lost and cannot be salvaged, it even more dramatically increases the inefficiency. The cost of producing this lost material is lumped together with the overhead to raise the cost of the whole operation.

A clean plant is usually safer. There are fewer accidents which are costly in both personnel and in production. A study by the U.S. Bureau of Mines on accidents states that a high percentage of accidents are conveyor related, and that for the most part, the victims were in a dangerous situation because of accumulated "DURT" from a conveyor.

A clean plant has better employee morale. People tend to feel proud that their place of work is a showplace. Who would brag about working in a sewer? A happy employee is more likely to give his or her best effort.

A clean operation seldom receives harassment from OSHA or MSHA. Why? Because it is obvious to these agencies that management is running a "tight ship"--doing its best to do things properly, efficiently, in everyone's best interest.

The problems that arise from "DURT" are obvious and costly. Then why are so many operations dirty? The answer to that is usually, the people who can cause things to happen in the plant are not aware of the "DURT" problem. With the increased use of computer systems, profit leaks will easily be spotted and brought to the attention of these managers. "DURT" will then be exposed for the costly problem it is.

"DURT" FROM CONVEYOR SYSTEMS

Studies have shown that the conveying system can be the predominate cause of fugitive material. An engineer for a major steel company studied their fugitive material problem and determined where it was coming from. His work showed that 80% escaped from the conveyor belts. The next step was to devise a program to eliminate this "DURT". The solution adopted was to install belt cleaners, but these cleaners quickly became inefficient, allowing fines to be carried back to become "DURT" piled along the conveyor way.

Finally, it was determined that servicing the belt cleaners weekly would control the problem. The maintenance cost on the belt cleaners rose by a factor of 10, but the service requirements for the other conveyor hardware fell dramatically. The savings in labor--labor no longer needed to clean up "DURT" along the beltway --was $1.5 million.

It is generally accepted that carryback on belt conveyors causes "DURT". But the quantity--just how much "DURT" there is--is not documented. What is needed is a method to measure the amount of carryback on a belt, and the amount of this material that becomes fugitive.

QUANTIFYING THE PROBLEM

I use a gauge of my own design to measure the amount. This gauge (FIG. 1) has a catching pan beneath a one inch wide scraping blade. When the blade is pressed against the underside of a moving belt, it scrapes the surface free of the particles which are clinging to it. By weighing the amount of material collected in a 10-second period as the average and multiply that weight by the width and speed of the belt, you can establish an accurate measurement of the material which is being carried back on the belt.

When I tell a client that a carryback of one gram per meter of belt length can result in over 2.7 tons per week of fugitive material from a single conveyor belt that is moving at 800 feet per minute, he's generally amazed.

Here's the calculation:

800 ft/minute = 243.48 m/minute

X 1 gram/m X 60 min/hr X 8 hours/shift X 6 days/wk

= 1,548 #/wk for one-shift per day of operation.

or 3,096 #/wk for two-shift per day operation.

or 4,644 #/wk for three-shift operation per day.

For round-the-clock, 7-days-per-week operation, the carryback would total 5,418 pounds or 2.709 tons.

Then I go out to their belt and measure their actual carryback and typically find it to be more in the neighborhood of five to ten grams per meter of belt length.

I invite you to take a spatula and scrape a meter of length clean from one of your typical belts. Weigh the material you've scraped and make your own calculation of the amount of material that is in a position to become fugitive and add to your "DURT" problem. For safety's sake, scrape from a belt that is not running. I believe you'll be surprised, scared and saddened by the amount of material you'll find.

Let's look at an example at what this kind of fugitive material can cost an operation. A barge unloading facility on the Ohio River incorporates a bucket type unloader that discharges to conveyor belts to bring the coal to their storage pile. So much of the material being unloaded escaped from the system and fell into the river during the barge unloading, that they had to dredge the unloading site three times per year in order to float a loaded barge into the slip. Each dredging cost $25,000, filled one or more 300-ton capacity barges. This coal was sold back to the station by the dredging company, which means they paid for this coal twice.

By plugging the leaks and eliminating carryback on the conveyor, the plant reduced the need for dredging to once per year. By reducing fugitive material, this facility created a direct savings of $86,000 per year from an investment of less than $15,000 in new equipment, sheet metal fabrication and controls.

Here's another example. A global mining company did a study directed at evaluating their various methods of materials handling. This study pointed out that the use of belt conveyors as a transportation system held particular advantages in terms of:

a. lower personnel costs

b. lower energy costs

c. lower maintenance costs

The study also indicated conveyors offered the potential for a higher degree of automation in the firm's material handling. But the study exposed a considerable profit drain through loss of material which became fugitive during its conveying.

The company's combined annual production of just two minerals, iron ore and coal, is over 60 million tons. A study of the handling at their port sites--without regard to their mines and process plants--indicated that material spillage amounts to approximately .05% of the total throughput. That means a material loss of 30,000 tons. If there was no fugitive material, there would be 30,000 tons of additional material to be sold.

The true costs added to their operation should not be limited to the loss of saleable material. In reality, this cost includes the expenses incurred cleaning up the lost material, as well as the cost of unscheduled downtime, which can stem from one of several "DURT" related causes, such as:

Build-up on idlers and pulleys, causing them to fail or misalign the belt.

A conveyor fire that shuts down the plant that is traceable to an idler bearing seized to a "DURT" build-up.

A health, safety, or insurance inspector judging the operation too dirty, and mandating a shutdown and clean-up.

The loss of services of a valuable employee injured in a "DURT"-related accident, such as a fall from a slippery catwalk, or getting caught in a pinch point during clean up work.

All these expenses--difficult, if not impossible, to tabulate--add to "The Price of DURT."

It is noteworthy that in their effort to reduce spillage and increase profit, this global mining firm instituted a plan directed at controlling the "DURT" from their belt conveyors, identified as the major source of their fugitive material. Let's take a closer look at conveyor systems, and the ways they cause "DURT".

CARRYBACK CONTROL AT THE HEAD PULLEY

The head pulley where the conveyor discharges is the single most important area of the conveyor to examine when your purpose is to control fugitive material. This is because the carryback on the belt accounts for the majority of fugitive material. Carryback can best be prevented with the installation of well-designed belt cleaners at the head pulley.

The problem comes because most often not enough room is provided in the design of the head frame and housing for an adequate belt cleaner system. And this is because designers do not take into account the sticky nature of the conveyed material when it is in its worst condition. Once the conveyor is designed, a belt cleaner specialist will be consulted for a recommendation on how to prevent carryback. But when the designer discovers that there is not sufficient room to accommodate the number of cleaners recommended, that recommendation will be compromised. The conveyor designers are not held accountable for the carryback specifications of the conveyor, so it is natural for them to "save money" and do less than the recommendation. Or they will find a belt cleaner supplier who will provide equipment that will fit into the system with the vague promise that it will do the job. This is the way it is, and the way it will continue to be, until specifications establish a quantitative amount of carryback that is allowable.

To analyze the material in its worst condition may take some doing, but believe me, it's worth the effort. Unless you do, you cannot have an efficient belt cleaner system. And if the system is inefficient and you get a bad run of material, the system will become a mess in a matter of hours, or even minutes. An efficient belt cleaner designed for worst case materials will be a case of overkill, under normal conditions. But when the material changes, you'll be prepared.

Another special consideration is the conveyor's dust suppression system. If it incorporates water sprays to wet conveyed material, it can accelerate the problems in belt cleaning. The fines become even more sticky, until they reach the point where they can overpower the cleaner. The designer should consider this and include a belt cleaning system that will handle this added carryback. In some cases, particularly if the system has only limited space at the head pulley, a scavenger conveyor (FIG. 2) may be required to return material into the system.

One of the keys to effective belt cleaning is the mating of blade to belt. It stands to reason the more perfect the blade mates with the belt, the better it will clean. Anything that makes it more difficult for the blades to stay in perfect contact as the belt moves must be considered as being undesirable. These factors would include out-of-round pulleys, or the poor choice or installation of lagging. Any pulsating vibration on the belt's surface can lower the cleaner's efficiency and adversely affect the life of the belt.

In any case, you will discover that it generally requires more than one device to properly clean the belt when the material is sticky. The location and type of device used is your decision. I like to use as a pre-cleaner a device I call a Doctor Blade because its location is at the very face of the head pulley.

Pre-cleaners

The position of the pre-cleaner blade in relation to the belt places it in a peeling position. (FIG. 3) With an angle of attack of less than 30°, this pre-cleaner blade cleans with very little tip pressure against the belt, resulting in low wear in both the blade and the belt surface. If this angle of attack were greater (a "scraping" position), greater pressure would be required to hold it in position against the onslaught of material.

This is an important consideration. Since the objective of a belt cleaner is to prolong the life of the conveyor components by removing fines, it would certainly be counter-productive if the cleaner itself contained the potential for damage to the belt. Lower blade pressure means that the device's tensioning system will be able to relieve its blade from the belt when a mechanical splice moves past the cleaning edge.

The Doctor Blade type cleaning device is designed to remove the majority of the material clinging to the belt right at the face, so these fines can join the main cargo discharged from the belt. This device acts as a lopper blade removing the heavy layer of carryback, leaving only a thin skim of fines to be removed by the secondary cleaner(s).

Secondary Cleaners

The positioning of the final cleaner is critical. The closer you remove the carryback to the discharge point, the lower your potential for problems with fines build-up on the dribble chute. I recommend a design that will allow the secondary cleaner blades to contact the belt as it leaves the head pulley. (FIG. 4)

For the secondary cleaner, I recommend a design that incorporates multiple blades approximately six inches wide that span the belt. Because this cleaner must be as efficient as possible, the blade must mate with the belt as flat as possible. There are peaks and valleys on the surface of the belt that the cleaner must adjust to instantly. The individual blades afford the best chance to effectively clean this moving surface.

The blades themselves should be of a hard material that resists the heat build-up from its scraping action against the belt's surface. Individual blades that are individually suspended have the best potential to remain in precise contact as the belt surface passes across the cleaning edge.

Again, the angle of attack of the blade to the belt is an important consideration. For this cleaner, I recommend a scraping angle, as blades in a peeling position are quickly honed to extreme sharpness. This raises the risk of an untrained or "in a hurry" adjustment while it will provide good cleaning, applies too much pressure to allow instant release from obstructions such as mechanical splices. The end result would be damage to the belt, splice, or the belt cleaner itself. Again, the cleaner and tensioning system must be designed to allow a splice to pass without damage.

Dribble Chutes

The build-up of fines on the dribble chute must be prevented. Often, a belt cleaner is judged to be inefficient, when in reality scraped off carryback fines have built up in the dribble chute, and encapsulated the cleaning device, preventing it from performing the job it was designed to do.

One way to solve the dribble chute build-up is to create a dynamic sub-floor in the chute. (FIG. 5) This is accomplished by mounting a sheet of smooth, low-friction, abrasive resistant plastic, such as UHMW Polyethylene, so the plastic is parallel to the chute floor but free to move. Use a "U" shaped piece of metal to support the sub-floor at one end; leave the other end unsupported. By mounting a vibrator to the free leg of the bracket, its vibrating action is transmitted through the leg and the plastic, providing dynamic action to prevent material build-up. Because this installation is not connected to the chute wall, there is no force applied to the structure to cause metal fatigue and cracking.

Belt Cleaner Maintenance

It is important to provide access to your belt cleaning devices. This will allow operations personnel to make inspections and provide appropriate service as needed, perhaps even several times a day on those occasions when the material is in its "worst case" condition. Access windows with easy to operate closure doors should be installed on both sides of the pulley, and in line with the axle of the belt cleaners.

You'll want to perform required adjustments in the shortest possible time, without the need for extended conveyor downtime. (MSHA does not permit service to be performed around a moving belt.) It would also be desirable if all adjustments and blade replacement could be performed with minimum tools. For this reason, I recommend the track mounted type of belt cleaner which permits an operator--even if he isn't a tool-carrying maintenance man--the ability to remove the cleaner and provide any necessary service during a brief conveyor outage.

CONTROLLING "DURT" AT TRANSFER POINTS

At the opposite end of the conveyor from where the conveyor belt delivers its cargo is another "DURT" producing area--the loading zone. Here, the culprit most often blamed for escaping material is the skirt rubber. This rubber is installed to provide a seal between the moving belt and the stationary steel which directs the material onto the belt. It would seem obvious that for this to be accomplished efficiently, the belt's line of travel should be as steady as possible. But this is not typically the way loading zones are designed. And even when designed properly, these transfer points are generally so poorly maintained that they develop unwanted dynamic action that defeats the skirt rubber's ability to seal effectively.

For example, a frequently encountered problem in the loading zone that contributes to a "DURT" problem is a conveyor design that permits the belt to rise off the idlers at start-up. Also, a belt that raises off the idlers when running empty can create the same type of sealing difficulties. In either case, in order to accommodate the different traveling lines of the belt, the design of the sealing system is compromised, and like all design compromises, suffers in effectiveness.

If a transfer station has the objective of being "DURT" free, or at least to allow as little "DURT" as possible, the designer should do whatever possible to keep the belt's travel line consistently straight. This is fundamental and must not be compromised. The loading of the belt with its cargo, the tensioning of the belt by the counterweight, the type, spacing and arrangement of idlers, the construction and shape of the belt in the loading zone, and the type of pulleys employed: all these factors influence the running line of the conveyor.

If a belt was supported by a flat table that prevented its moving in any direction except in a straight line, it would be easier to place skirt rubber against it and expect to maintain a constant seal while the belt was in a traveling mode. Loaded or unloaded, stopped or in motion, the belt's position would be the same. The conveyor system designer should attempt to achieve a support system for the belt that is similar to the table's ability to keep the running line true.

Belt Support and Idlers

The idler spacing through the loading zone must be as close as necessary to keep the belt from sagging between idlers under load (FIG. 6) Any measurable belt sag--even one that is not apparent to the naked eye--is enough to permit fines to start a grinding action that produces wear on rubber skirt and belt surface and allows the escape of materials to the environment.

Idlers should be aligned with care and matched so as not to produce humps or valleys in the belt. Idlers should be checked for concentricity, for the more they are out of round, the greater tendency for the belt to dance.

Impact idlers with a rubber disc for the rolls are more likely to be out of round than the regular steel idlers. Of course, the conveyor designer should consider that when the idlers are 8 to 10 inches apart, they share the impact load to a much greater degree than when placed at a 12 to 18 inch interval. Consequently, rubber impact idlers may not be necessary unless you are dealing with large lumps--say, 2-1/2 inches or larger.

Because of their design intention to swing and reshape the belt, catenary idlers can create a sealing problem. They can be a valuable aid in handling severe impact and off center loading problems, but the dynamics they add to the sealing surface must be considered in engineering the conveyor system.

Slider Beds and Cradles

So important is the flat table concept to good transfer point sealing that many designers are now using slider beds or cradles (FIG. 7) instead of idlers in the transfer point. This is accomplished by using bars that feature the ability to cushion impact through the incorporation of a shock absorbing elastomer. These bars are then covered with a surface layer of plastic with low friction and wear-resistant properties. A cradle made with these bars may add slightly to the friction of the conveyor and, hence, to its power requirements. But this marginal increase in energy consumption is more than offset in the profit picture through the elimination of the expenses for the clean up of skirt leakage and the unexpected downtime necessary for idler maintenance and belt replacement.

In areas of low impact, a single slider bar can be installed on each side of the conveyor belt providing a side rail effect which helps support the belt. (FIG. 8) This technique provides a smooth, low friction, fixed-in-place surface onto which the skirt rubber can form an effective seal.

Belt Troughing and Transition

Introducing the load before the belt is fully troughed can also create a sealing problem. As the belt comes around the tail pulley and is bent from a flat position into a troughed configuration by the transition idlers, the belt edge, where the rubber skirting is to contact the belt, is constantly changing and difficult to seal. Consequently, the distance from the tail pulley to the loading point should be far enough so the belt is fully troughed when it enters the skirted area. A longer design allows more room for the belt to assume its fully troughed position before the introduction of the load and the application of a skirt seal.

The tail pulley itself needs to be considered. If it builds up with material, it will cause the belt to dance. This action is a vibration of the belt surface that makes it difficult to maintain an effective seal. Any condition that allows the belt to dance should be considered detrimental to sealing, and therefore, contributing to "DURT". As an example, consider an idler that, due to a bearing problem, has stopped rotating and allows the rubber to be worn flat by the moving belt. The problem may have been a piece of foreign matter that fouled the bearings. If it gets flushed out with the next lubrication, the idler will again rotate. But the idler's flattened surface will cause the belt to dance, creating a sealing problem. Quick replacement of idlers that are causing the belt to vibrate are a priority for maintaining a clean transfer point. For this reason, an idler system or impact bar system should have features that make it easy for the maintenance people to perform a quick replacement of faulty components.

The dust curtain where the belt leaves the loading zone should be maintained so it allows minimum outside air into the transfer point. Dust collection equipment in the transfer point has all it can handle gathering up dust-laden air. The more free air introduced, the less efficient the dust collection system will be. This is also true for the dust curtain positioned at the discharge chute area.

The conveyor system designer should take special precautions when a transfer point is beneath a crusher. The downward draft created by a crusher into the transfer chute enclosure will be difficult to seal. The micro-mesh particles in the material will tend to blow out every crack in this enclosure. Even the hole left by a missing bolt will account for tons of dust being released into the environment in a surprisingly short period. Extra ducting to the dust collector--with special attention to providing a positive air pressure into the enclosure--will make it possible to have effective skirting and a dust free transfer point.

Skirting systems

There are a number of rubber skirting systems available. Again, my preference is for a system that is simple to install and maintain. This points toward that system which is composed of short rubber blocks which allow for individual adjustment and replacement (FIG. 9), while at the same time allowing the system to better cope with the compound curves frequently presented in a belt line as it is troughed. This system provides for adjustment with minimum tools or fuss and has the added benefit of allowing replacement while the belt is running.

CONCLUSION

Ease of maintenance is essential to the concept of having a "DURT"-free conveying system. Management may be committed to a "clean plant" philosophy, but the need to keep production up will cause "DURT" considerations to be compromised. Maintenance crews cannot call out for the belt to be down for a lengthy period in order to change out an idler or belt cleaner. That would be too costly to production efficiency. Consequently, all items that serve as components in the fugitive material control system should be designed to feature easy maintenance and replacement.

"DURT" is a costly problem. It decreases production efficiency, increases the costs of maintenance, poses a safety risk, and acts as a lightening rod for the attention of environmentalists and regulators. By taking positive steps to control fugitive materials and eliminate the sources of "DURT," you improve your operation--in safety, in the quality of work life, and in production efficiency. These improvements will appear in the balance sheet. Cut the "Price of DURT." THINK CLEAN!

##

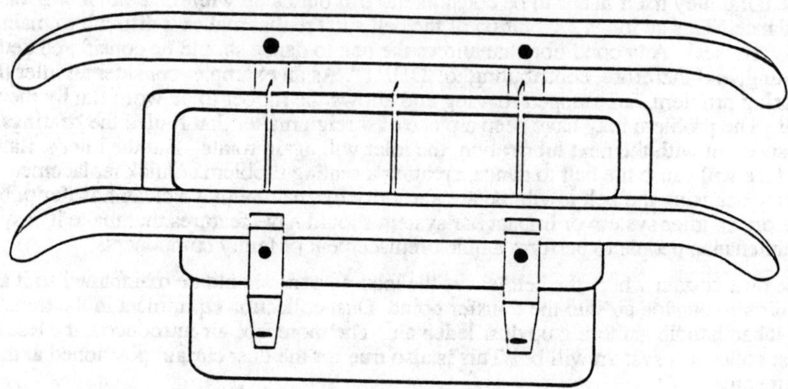

FIG. 1
A "DURT GAUGE" composed of one-inch wide scraping blades and a catching pan is used for measuring carryback.

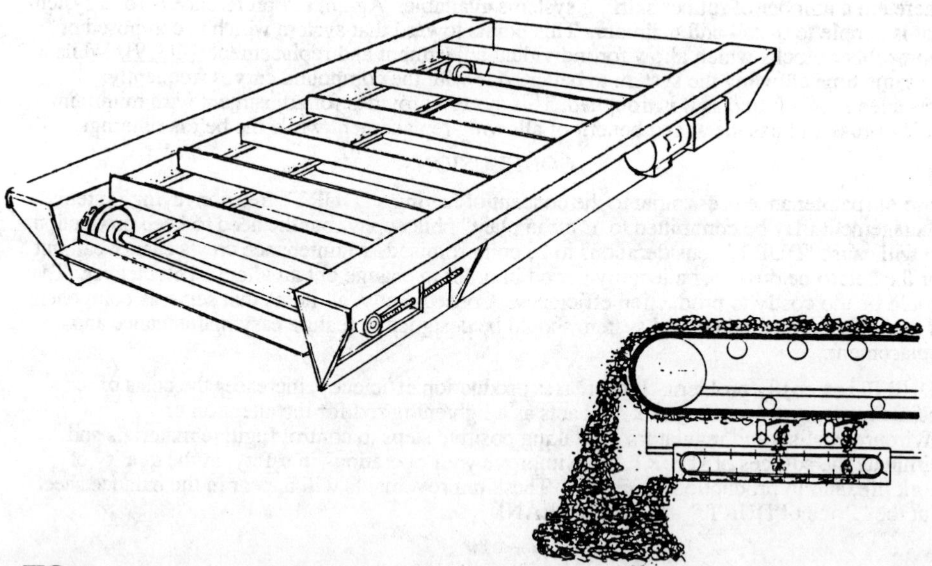

FIG. 2
A scavenger conveyor is useful to return fugitive material to the main conveyor system, especially where room is limited at the head pulley.

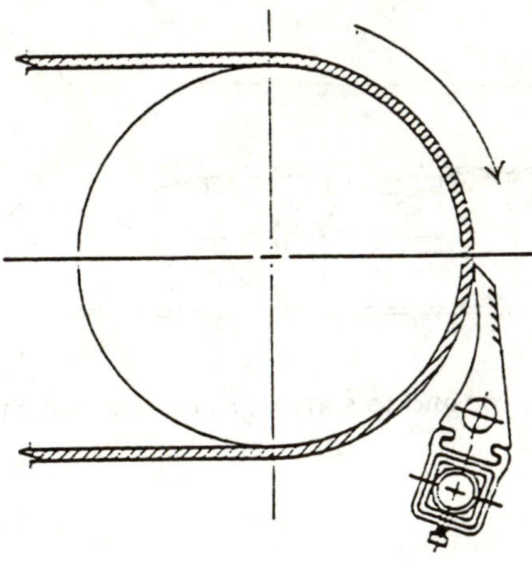

FIG. 3
A"Doctor Blade" pre-cleaner is installed in a peeling position on the face of the head pulley. to remove the majority of carryback.

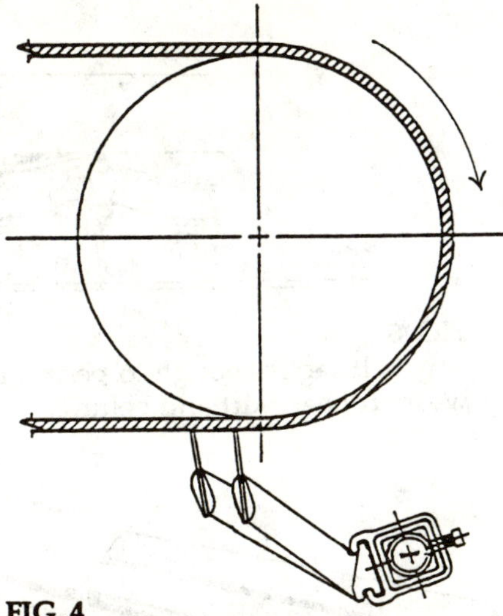

FIG. 4
The Secondary Cleaner is positioned just as the belt is leaving the head pulley to remove residual fines.

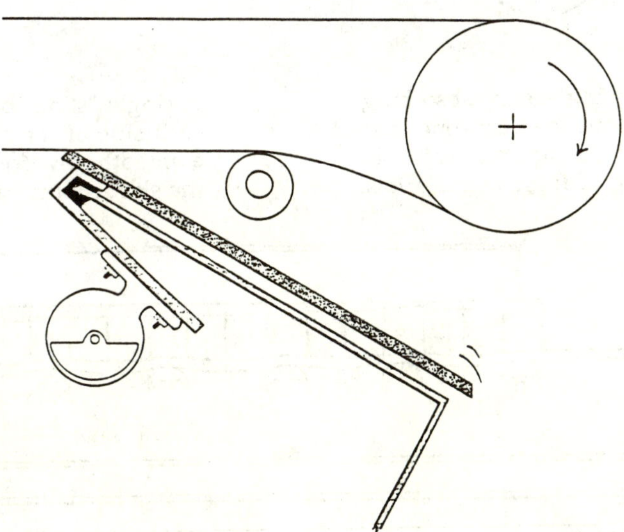

FIG. 5
To keep material moving through dribble chutes, create a dynamic subfloor in the chute using a low friction plastic and a vibrator on a isolation mount.

FIG. 6
Any belt sag is enough to permit fines to escape and to start a grinding action that wears rubber skirt and belting.

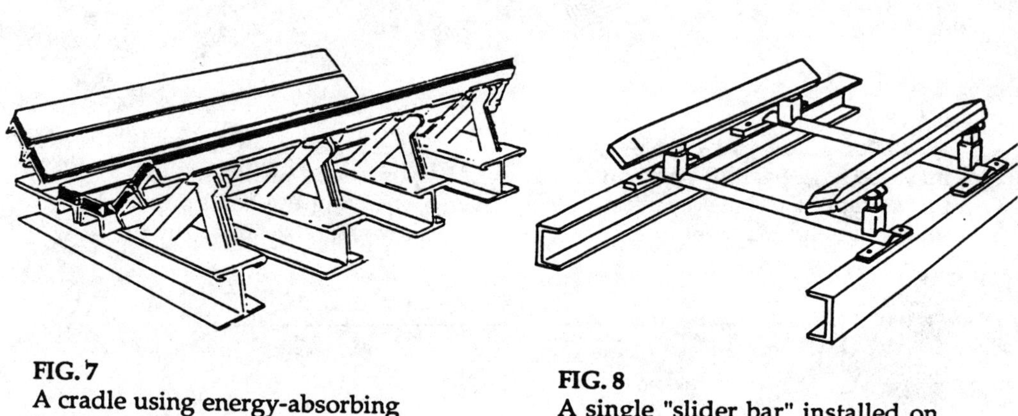

FIG. 7
A cradle using energy-absorbing bars cushions loading zone impact to prevent damage and control the release of fugitive materials.

FIG. 8
A single "slider bar" installed on each side of the conveyor provides a smooth, low friction surface which the skirt rubber can effectively seal.

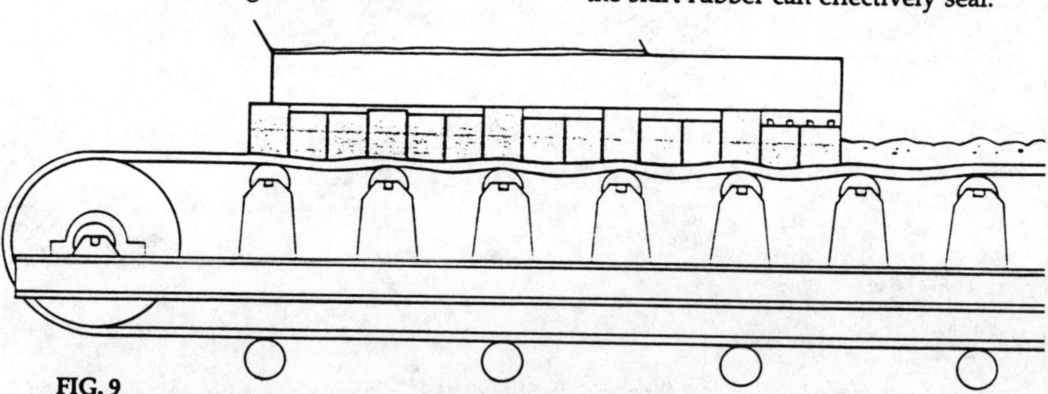

FIG. 9
Skirting systems composed of short rubber blocks allow precise individual adjustment to accommodate variations in belt travel.

Author Index